ASTROPHOTOGRAPHY II
Featuring the Techniques
of the European Amateur

Patrick Martinez

ASTROPHOTOGRAPHY II
Featuring the Techniques of the European Amateur
Patrick Martinez

Translated from the French by

Charles Miller

Published by:

Willmann–Bell, Inc.
P.O. Box 35025
Richmond, Virginia 23235 ☎ (804)
United States of America 320-7016

Publishers and Booksellers

Serving Astronomers Worldwide
Since 1973

Previous Editions: French Language © 1983 by Patrick Martinez

Library of Congress Cataloging-in-Publication Data

Martinez, Patrick.
    Astrophotography II.

    Translation of: Astrophotographie.
    Includes index.
    1. Astronomical photography.   2. Astronomy—
    Amateurs' manuals.   I. Title.
QB121.M3713   1987   522'.63   87-2143
ISBN 0-943396-13-1

Printed in the United States of America

        91 92 93 94 95 10 9 8 7 6 5 4 3 2

PREFACE TO THE ENGLISH EDITION

During the 1830's Louis Daguerre in France and Fox Talbot in England independently made pictures by the action of light on a sensitized material. The cornerstone of modern photography was set in January, 1839, when the work of these experimenters was publicly announced. The first astronomical "photograph," a daguerreotype of the Moon by New York physician John W. Draper, was obtained the following year. But it wasn't until the 1880's, a mere 100 years ago, that photography advanced enough to produce practical records of the starry sky. Before the turn of the century, pioneering astronomical photographers like Edward Emerson Barnard and Isaac Roberts were peering across the cosmic void with their cameras. With a few hours' exposure and a splash of developer they were revealing a universe that had been hidden from sky gazers for thousands of years.

Coupled with the telescope and spectroscope, the photographic plate was a powerful tool. It helped map the Milky Way and led to the realization that our solar system was in a distant corner of a vast pinwheel of countless suns. Moreover, it revealed millions of other galaxies in an expanding universe. Along with the science in each photograph came a richly detailed "picture." Just as the cameras of Ansel Adams and Alfred Eisenstaedt showed us a world many might not have otherwise seen, those of the astronomers brought us the universe. Beautiful pictures of brilliant star clusters, turbulent clouds of gas and dust, and distant galaxies were set before our eyes.

Astronomers today, however, have largely replaced the relatively inefficient photographic plate with electronic charged-coupled devices (CCD's). Although the science now pours forth at an ever greater rate, CCD images lack the beauty of a simple photograph. This may change, but for the present it is the amateur who must tap the potential of astronomical photography. We are the guardians of an "old" technology. Whether it's the sudden flash of a meteor across the sky, Orion suspended above a cold winter landscape, or the glow of some far-away galaxy, the most aesthetically pleasing images now come mainly from the cameras of amateurs.

I took my first night-sky photograph more than 25 years ago with my mother's folding camera. I've taken thousands more since. Along the way I've witnessed great changes in the field of astrophotography. Not only have I seen my own techniques and results improve, but, as an editor of *Sky and Telescope* Magazine for more than a decade, I've followed the work of hundreds of other photographers, who have submitted their pictures to the magazine.

In the early 1960's, the first issue of *Sky & Telescope* that I read had but a single ad for an astrophotography accessory — an electronic drive corrector. Magazines today contain pages of gadgets to help astrophotographers. No longer is it a prerequisite for an astrophotographer to be a telescope maker. In the early 1970's, Tom Johnson offered deep-sky photography to the masses with the introduction of an astrophotography system built around his Celestron 8. Other systems followed. Modern high-speed films and simple camera mountings further encouraged amateurs to try their hand at astrophotography. As a result, there are more people actively photographing the night sky now than ever before.

Recently a number of books have appeared for the novice. Like these, Martinez' work includes material necessary for the beginner. But its scope is broader, covering such subjects as films, developers, filters, and optical systems in greater detail than other books. This is important for those wishing to hone their skills, because astrophotography remains one of the most technically oriented fields of photography. We can't change our viewing angle or dictate how the subject is illuminated. If we are to improve our pictures, we must improve our techniques. Martinez tells us how — whether it is the Sun, Moon, planets, or the universe beyond that we wish to photograph.

Dennis di Cicco

November 30, 1986
Sudbury, Massachusetts

PREFACE TO THE FRENCH EDITION

It is a rare amateur astronomer who has never tried at one time or another to delve into astrophotography.

He wants to capture an infrequent phenomenon, for example, such as an eclipse. Or record objects scarcely discernible by the naked eye, but which will register perfectly on film after a sufficiently long exposure. Or again, he aspires to assemble an album to use as a basis of reference for future comparison. Motivations abound, and the results of the first attempts are usually so gratifying that the amateur soon wants to know more about photographic techniques.

It is in order to help in the acquisition of this technique that Patrick Martinez has written this book. He addresses both the beginner—whom he will guide through the preliminary steps—and the seasoned amateur, who will learn how to choose the method best adapted to a particular observation.

But fortunately this book is not reduceable to a collection of recipes to be followed blindly in order to avoid failure. On the contrary, the approach is to analyze in detail the problems posed by each type of observation—solar, planetary, stellar—and to lead readers to finding for themselves the solution to adopt, after taking into account the methods at their disposal.

And these methods can be quite modest, as P. Martinez correctly insists. To begin astrophotography, you need not have a top-of-the-line Schmidt-Cassegrain scope with a cold camera and an equatorial mounting driven by a microprocessor. By way of illustration, you will forgive me for mentioning my own debut as a young amateur astronomer and my immense satisfaction when I obtained my first pictures of the moon with a refractor built around an objective salvaged from a tank periscope and using a plywood camera.

Today amateurs enjoy much more privilege, since they can equip themselves with excellent material on a relatively modest budget. In addition, film emulsions have greatly improved, color has become more accessible, hypersensitization processes have been simplified. Still, one must keep abreast of all these possibilities and that is precisely the purpose of this book. The optical combinations, films, and photographic treatments presented and discussed have all been tried by P. Martinez himself. He has placed the experience he has accumulated over long years of observation within the reach of everyone.

There is nothing more for me to say, except to wish you much enjoyment reading this book, followed by a great deal of pleasure in your photographic explorations of our universe!

—Jean-Paul Zahn
Director of Observatories
of Pic-du-Midi and Toulouse

INTRODUCTION

A few years ago I had a group of people curious about the sky visit the observatory of the Société d'Astronomie Populaire. That evening the telescope was pointed at the moon and I invited my visitors to contemplate our satellite. In less time than it takes to say it, one of them "unsheathed" the camera slung over his shoulder, aligned it vaguely behind the eyepiece and clicked the shutter.

I never learned the result of this improvised picture, but one can imagine it without much trouble. This anecdote is a small caricature of the simplistic conception many amateurs have of astrophotography. Quite to the contrary, however, astrophotography is a discipline in which art and improvisation must give way to technique.

This is the first message this book wants to teach. While eschewing the driness of a mathematical treatise, it tries to show readers that rigorous reasoning lets them achieve optimal shooting conditions fairly rapidly, thus avoiding useless trial and error.

The second characteristic of this book is that it provides methods rather than isolated results. Indeed, optimal solutions depend upon the equipment possessed by the individual and are likely to evolve fairly rapidly in time with technological progress. The empirical "rules of thumb," which to the beginner can seem like easy solutions, often prove ineffective because used inappropriately without the necessary reflection are always sterile for amateurs wishing to understand the phenomena confronting them.

Usually a work with such aims is intended for the erudite and proves inaccessible to beginners. This book, however, strives to be accessible to the general public. Thus, elementary concepts about telescopes are explained for beginners in the first chapter, and important concepts like the resolving power of films are treated from several angles—on the one hand with an intuitive approach accessible to all readers, on the other hand in more rigorous terms more apt to satisfy the specialist.

The book as a whole constitutes a calculated progression along the path each astrophotographer must follow.

The first chapter, which the reader may skip, contains essentially elementary principles for the beginner, as well as a few points on optics where purists will find the justification for setups and calculations encountered later on.

The second chapter, on photography without a telescope, addresses beginners especially, since they are the first to practice it. Hence, it does not cover certain theoretical considerations on the use of films or certain special techniques, which are treated later on.

Since camera lenses in many cases prove insufficient for astronomy, the use of a telescope as a light collector becomes necessary. Chapter III, therefore, describes setups permitting the use of a telescope in astrophotography.

After the telescope, the other fundamental tool is the film. More demanding than the traditional photographer, the astronomer must know a film's characteristics, its performance capabilities, and its limits. These are explained in Chapter IV, which concludes with the presentation of a constraint—called the "luminosity equation"—relating all the shooting parameters to one another.

This constraint implies the existence of a optimum choice of parameters. Now, the search for this optimum must differ according to whether one is conducting planetary, solar, or faint-object photography. The three following chapters are therefore devoted to the discussion of optimal conditions for these three major categories.

After discussing the methods (optical setups, films) and their combinations in planetary, solar, or faint-object photography, we must not forget that astrophotography requires the mastery of a certain number of techniques which, though not entirely specific to this discipline (focusing, stationing, etc.) require here more than elsewhere much rigorousness on the operator's part. These techniques, developed in Chapter VIII, are sometimes neglected by amateurs—often one need not look elsewhere for the source of many setbacks.

Finally, the last chapter touches upon certain notions likely to help the amateur in the selection of material adapted to astrophotography.

The author wishes to thank everyone who helped in the making of this book, especially: Jean-Paul Zahn, who kindly agreed to write the preface to the French edition, my English translator Charles Miller and Dennis di Cicco for his preface to this English edition. Also, Jacques Barthes, who did the drawings for the book; Alain Maury, for his valuable collaboration on everything concerning photographic film; Honoré Arioli, specialist in solar eclipses; Eliane Lay, who typed the manuscript; Philippe Alexandre, for his effective participation in correcting and proofing the text; all the amateur astronomers who furnished their photographic work; and finally, all the members of the S.A.P. who saw to certain, sometimes fastidious material tasks.

Patrick Martinez

TABLE OF CONTENTS

The picture on the opposite page was taken by the Author and is of the Eta Carina, a nebular variable star. The telescope was a 12-inch Newtonian, F/D = 6, on Kodak 2415 gas hypersensitized film and exposed for 45 minutes.

CHAPTER I
REVIEW

I.1 — CELESTIAL OBJECTS

Before discussing the techniques of astrophotography, it is a good idea to review the objects that we will have occasion to photograph.

Obviously the luminosity of heavenly bodies is going to play a large part. Every photographer knows that he must adapt himself to his environment: in broad daylight he will use a "slow" film with a short exposure time and a higher f-stop; conversely, indoors, if he does not have flash, he will need a "fast" film, a lens opened very wide, and probably a support which allows him to make relatively long exposures.

Note, however, that the luminosity of the objects that interest the conventional photographer (souvenir photos, family shots, travel pictures, etc.) can vary from 1 to 1000, whereas a factor of several billion occurs between the sun and a nebula! We see right away that there must be not just a single technique for astronomical photography, but several.

Nevertheless, if we analyze the heavenly bodies, we discover that we can arrange them by their luminosity into three large families, to which there correspond three types of photography:

- solar photography: the sun,
- planetary photography: the moon, Mercury, Venus, Mars, Jupiter, Saturn,
- faint-object photography: stars, distant planets, asteroids, comets, nebulae, clusters, galaxies.

Here is how we can differentiate between these categories on a technical level. In solar photography our telescope receives too much light—it is the only time when we will have occasion to use filters or other optical systems with the sole aim of reducing the luminous intensity reaching the film. This operation, which seems easier than it really is, will be our principal concern. In planetary photography the problem is reversed: our telescope gathers too little light and this fact will be one of our main preoccupations in our choice of film, magnification, exposure, and filters. Nevertheless, the luminosity of the planets enables us to seek a compromise when choosing from these parameters to obtain the finest detail. On the other hand, this ceases to be possible for the celestial objects found in the third category: here the photographic system should be designed to give priority to capturing their extreme faintness.

In astronomical photography the visible grandeur of the heavens is not a very important criterion. The conventional photographer adapts the focal length of his lens in order to fit the subject to the format of the film; thus, a wide-angle lens (that is, one of short focal length) is indispensable for capturing all the furniture of a room on the same frame. In astronomy most heavenly bodies are so small that the problem of field does not apply to them. Just imagine, for example, that it takes a 100 meter focal length for Jupiter to fill a 35mm frame! The choice of focal length, therefore, will be exclusively conditioned by the desire to have an image that is sufficiently large and, at the same time, sufficiently bright.

This rule has essentially two exceptions among luminous objects: the sun and the moon. If you want one of these bodies to appear in its entirely on the same negative, you have to make sure not to exceed about 2000mm in focal length (for 35mm format), consequently losing perhaps the smallest details in the grain of the film. Faint objects do exist whose visible dimensions equal or surpass those of the moon or the sun (large nebulae, extensive star fields, large comets, etc.) but the use of a short focal length is generally not an additional constraint in those cases, for it is also necessary to obtain an image of sufficient brightness.

Photograph 1-1. *On the opposite page is a photo showing many faint objects: the stars in the constellation Cygnus (Deneb is at the center), the Milky Way, the nebula NGC 7000 (under Deneb), the satellite Skylab (long streak) crossing a meteor (short streak). Taken by Paul Moithy on Tri X film.*

I.2 — PRINCIPLE OF THE CAMERA

The experiment we now describe is easy to carry out and requires only a simple magnifying glass. Nevertheless, it helps us understand how most cameras work.

Photograph 1-2. Magnifying glass across a light source.

In photograph 1-2 we have placed a magnifying glass across from a light source — in this case a window inside a room. We have placed a sheet of paper behind the magnifying glass. At some point between the magnifying glass and the paper (about 10cm or 4") we see an image of the window form on the paper. Notice that this image is reversed: the top represents the bottom and the right side the left side of the object.

Now, let's replace the paper with a light-sensitive surface which will retain the image it receives. In this case we have to protect the surface (let's call it "photographic film") from any parasitic lights, so we enclose the film in a black box where the lens (magnifying glass) consitutes the only opening. But in order to be correctly exposed, the film must receive the light for a precise period of time, hence the necessity of placing a shutter behind the lens. We have just constructed a camera made up of two essential parts: the body (film support, chamber, shutter) and the objective (lens).

Notice finally that if the distance between the lens and the paper is not correct, the photograph is blurred. Every camera, therefore, must have an adjustment for this distance—that is, it needs a "focus."

I.3 — REFRACTORS AND REFLECTORS

I.3.1 — PRINCIPLE OF THE REFRACTOR

Let's return to the previous experiment: the sheet of paper carries the image of the window formed by the magnifying glass. If this paper is transparent enough (tracing paper) we can see the image by standing behind it.

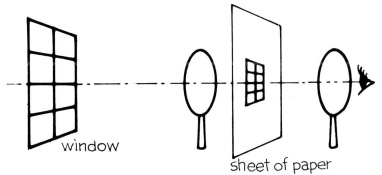

window

sheet of paper

Figure 1-1.

So, we can use a second magnifying glass to enlarge the image (Figure 1-1). Notice that the image seen through the second glass is still reversed and that it is sharp only for a precise distance between the paper and the second glass.

Now, let's remove the tracing paper. Through both the magnifying glasses we still perceive the same image of the window, whereas at the point where the sheet was we no longer see anything. Thus, in order for the image formed by the first glass to be accessible to our eye, there must be a material support, but the second glass can replace that support. Notice that the two magnifying glasses should be aligned as well as possible and that the image seen through them is sharp only for a precise distance between them.

In assembling this apparatus, we have built a refractor, though a rudimentary one to be sure.

In a refractor the first glass—called the *objective*—is generally made up of two, sometimes three, lenses placed or cemented together. This is done to eliminate a certain number of aberrations.

The role of the second glass is taken by the *eyepiece*. Because of the never-ending search for the best possible image quality, an eyepiece can consist of two, three, four, or even five lenses.

These optical pieces are fixed to two tubes, the objective tube and the eyepiece tube, which can slide one into the other to permit focusing.

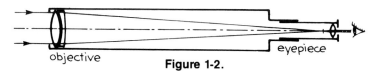

objective

eyepiece

Figure 1-2.

As shown in Figure 1-2, the objective and the eyepiece are of different sizes. Like camera lenses, the eyepiece and objective of a refractor are characterized by a focal length, which we will denote F for the objective and f for the eyepiece. It is because F is so much larger than f that a refractor enlarges the observed objects; in fact, its magnifying power M is given by the relation

$$M = F/f.$$

We shall discuss all this in detail in the Section 1.4 "Elements of Optics."

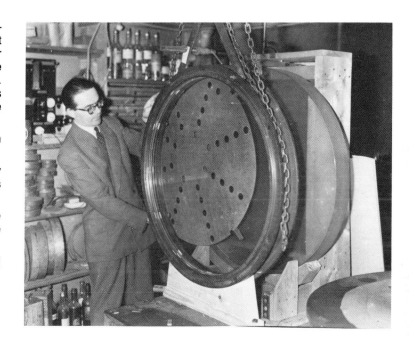

Photograph 1-3. *At one time refractors were the dominant telescope. Here, the famous French optician Jean Texereau, author of* How to Make a Telescope *adjusts a Hartmann test screen between the 806mm Henry Brothers objective of the Meudon Observatory and a test flat made by Couder.*

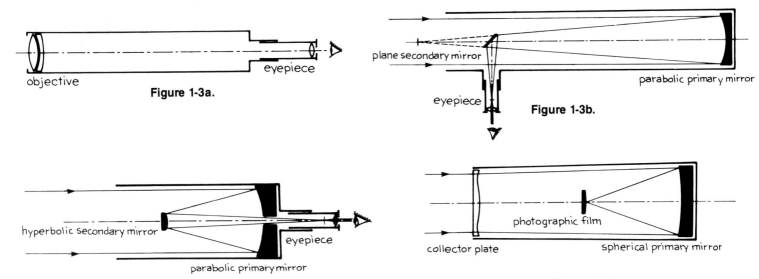

objective

Figure 1-3a.

plane secondary mirror

eyepiece

parabolic primary mirror

Figure 1-3b.

hyperbolic secondary mirror

eyepiece

parabolic primary mirror

Figure 1-3c.

collector plate

photographic film

spherical primary mirror

Figure 1-3d.

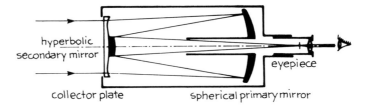

hyperbolic secondary mirror

collector plate

eyepiece

spherical primary mirror

Figure 1-3e.

I.3.2. — PRINCIPLE OF THE REFLECTOR

Like the refractor, the reflector requires an objective and an eyepiece. The objective, however, no longer consists of a pair of lenses, but of a concave mirror (usually parabolic) aluminized on the front surface. Properly shaped, this mirror plays exactly the same role as the objective in a refractor: it forms a "focal" image observable with the aid of the eyepiece.

The path of the light rays is shown for a refractor, a Newtonian reflector, and a Cassegrain reflector, in Figures 1-3a, b and c.

Note the presence in the reflectors of a secondary mirror, either flat or hyperbolic, in a clearly practical location. In fact, it would hardly be suitable to let the focal image be formed at the entrance of the tube: the eyepiece and the observer's head would create such a shadow on the primary mirror that observation would be rendered impossible. Therefore, we use a secondary mirror whose obstruction is relatively small to direct the image formed by the primary mirror either to the side of the tube (Newtonian reflector) or behind the tube through a hole cut in the center of the parabolic mirror (Cassegrain reflector).

Even if the parabolic mirrors in reflectors present an ideal image on their principal axis, considerable aberrations appear at the edges of the field. To avoid this problem, Schmidt reflectors have a spherical objective mirror with imaging of constant quality over the entire field (Figure 1-3d). The entrance of the tube is covered with a specially shaped window, or "correcting plate," to reduce aberrations caused by the fact that the primary mirror is spherical and not parabolic.

Schmidt telescopes, with their large, bright fields, are reserved for faint-object photography.

The Schmidt-Cassegrain system combines the spherical mirror and the correcting plate of the Schmidt reflector with the secondary hyperbolic mirror of the Cassegrain (Figure 1-3e). It is widely used by amateurs, especially because of its compactness.

The objective mirror of a reflector (or the combination of two mirrors in the case of a Cassegrain) is also characterized by a focal length F: the magnifying power furnished by an eyepiece of focal length f is still $M = F/f$. Every photographic setup that we consider consists of improving the focal image furnished by the objective, but it makes no difference in principle whether that image is supplied by a refractor or a reflector. Therefore, everything we explain henceforth can be applied indiscriminately (except where specifically noted) to either a refractor or a reflector, though the diagrams will picture refractors, which are simpler to draw.

I.3.3. — RESOLVING POWER—OBJECTIVE DIAMETER

The resolving power of a telescope is the value of the angular diameter (see Section I.4.2.) of the smallest discernible detail; therefore, it is a characteristic of the quality of the image obtained. Now, we shall see that the resolving power of an instrument is determined by the diameter D of its objective by the relation.

$$r.p. = 120/D,$$

where the resolving power is expressed in seconds of arc and the diameter in millimeters, or where the diameter is in inches:

$$r.p. = 4.56/D.$$

The greater the diameter of the objective, the better one can see small details.

Note that the relation we have just set forth gives the theoretical resolving power. This will not be attained unless the optical quality is excellent and the atmosphere is perfectly stable.

Here are some interesting orders of magnitude:

- diameter of the moon or sun: 30'
- resolving power of the eye: 1'
- maximum diameter of Jupiter: 48"
- theoretical resolving power of a 4.7-inch telescope: 1"
  —of a 47-inch telescope: 0.1"

This table shows us why Jupiter appears to the naked eye only as a point, whereas the moon has a perceptible apparent surface on which we can see details. On the other hand, with a 5-inch telescope under good conditions we can discern details on Jupiter whose size reaches about a fiftieth of the angular diameter of the planet.

Moreover, the amount of light collected by the objective (hence the brightness of the resulting image) is directly related to the diameter of the objective.

We see then that the diameter seems to be the essential characteristic of a telescope—it is what determines the instrument's capabilities.

The focal length and the type (refractor or reflector) are only secondary characteristics.

I.3.4. — MOUNTINGS

To be able to aim at any region of the sky, the telescope must maneuver around two axes.

I.3.4.1. — ALTAZIMUTH MOUNTING

The simplest idea is shown in Figure 1-4 and consists of using a vertical and a horizontal axis, like the turret of a cannon. This is an altazimuth mounting.

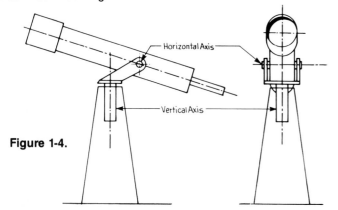

Horizontal Axis

Vertical Axis

Figure 1-4.

Although such a mounting permits easy aiming at a celestial body, it is less effective for following its movement. In fact, for the observer the rotation of the Earth about its polar axis causes an apparent rotation of the sky around the same axis in the opposite direction (from East to West) at a speed of one rotation every 24 hours. We are used to seeing the sun rise in the East and set in the West. The same holds true for the stars, and even though they seem immovable when we fix them with the naked eye, their displacement is quite noticeable in the field of a telescope magnifying some tens or hundreds of times.

Figure 1-5 shows the displacement in the course of one night of an object near the celestial equator. In order to follow this movement with an altazimuth mounting, you have to work both axes simultaneously and with varying rotation speeds, which is rather difficult to achieve.

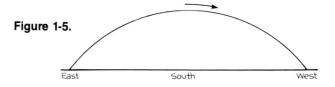

Figure 1-5.

East   South   West

Another serious drawback of the altazimuth mounting for long-exposure photography is the rotation of the field. Let's suppose we are photographing two stars A and B having the same declination and lying close to the equator (Figure 1-6).

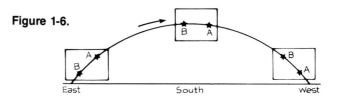

Figure 1-6.

East   South   West

The squares represent the field of the photographic plate. Since the telescope turns about a vertical axis, the edges of the field remain parallel and perpendicular to the horizon. By contrast, when you point toward the East, star A is above star B, and with their gradual displacement in the sky, star B passes above star A. All this happens as if the segment AB, and all the other objects photographed as well, had turned at a certain angle around the center of the photograph: the stars are no longer represent by points, but by arcs of circle.

The simplest solution to all these problems is provided by an equatorial mounting.

I.3.4.2. — EQUATORIAL MOUNTINGS

I.3.4.2.1. — Principle

Let's imagine that we incline the vertical axis of an altazimuth mounting so that it is parallel to the terrestrial poles. Note that in doing this the axis of our mounting is now directed toward the North Celestial Pole (the South Pole if we were in the Southern Hemisphere) and that its horizontal plane makes an angle equal to the latitude of the site. This axis, pointing toward the pole, is called the polar axis, and we have just "invented" a new type of mounting called an *equatorial mounting* (Figure 1-8).

Figure 1-7.

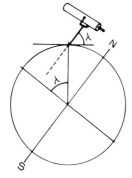

The polar axis is also called the axis of right ascension and the other axis the axis of declination. By turning the telescope about the former, in fact, you can vary the right ascension of the point at which the instrument is aimed. The latter axis permits you to vary the declination. Right ascension and declination are the two "equatorial" coordinates of heavenly bodies.

Figure 1-8.

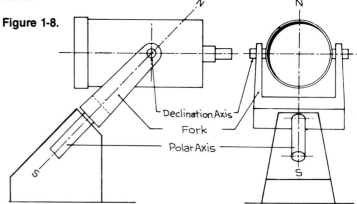

We can thus point to any region of the sky with the two axes of an equatorial mounting. Suppose now that the telescope is directed at a star and that we turn it slowly, because of the 24-hour rotation, around its polar axis from East to West. This movement is exactly opposite the Earth's; therefore, it is identical to the apparent motion of the sky. The target star thus remains fixed in the field of the telescope, and we have at the same time eliminated the rotation of the field. To accomplish this, we have had to make the telescope move about a single axis at a constant speed.

This movement can be created easily with a single motor at constant speed. Thus, it can be regular enough to permit photography without noticeable blurring. Apart from certain photographs of the sun (very short exposure) and possibly wide-field stellar photographs (hence faint resolution) a motorized equatorial mounting is indispensable to those aspiring for good results in astrophotography.

I.3.4.2.2. — The principal types of equatorial mounting

Having to mount the entire telescope on an inclined axis such as 45° poses some mechanical problems. In fact, every "cantilever" hampers the smooth working of the moving parts and is liable to produce vibrations. On the other hand, the mounting should not be too heavy if it is to be portable and, if possible, it should allow the telescope to be aimed in every direction.

This rigidity/encumbrance dilemma has given birth to various types of equatorial mountings. Here we describe the principal ones, classed in order of increasing rigidity.

German mounting (Figure 1-4a). The counterweight serves to place the center of gravity on the polar axis and thus balance the mounting, which is absolutely necessary in order to limit vibrations and permit a regular sweeping motion with a small-power motor.

Simple English mounting (Figure 1-4b). The adjunction of a North support permits better stabilizing of the whole.

Fork mounting (Figure 1-4c). In order to keep from limiting too much the field in the polar region, the tube of the instrument must be able to pivot about the axis of declination without touching the base of the mounting. In addition, to avoid the vibrations caused by the arms of the fork being too long, the instrument's center of gravity (through which the axis of declination passes) must be close to the back of the tube. These conditions exclude this type of mounting for refractors (among the rare exceptions is the solar refractor at Pic du Midi, but that instrument never points to a declination greater than 25°. Thus, one sees these mountings used with Newtonian or Cassegrain reflectors, though with the latter observation of the polar region is sometimes difficult.

English yoke or cradle-type mount (Figure 1-4d). This is without doubt the heaviest, most encumbering mounting, yet the most stable. Its principal defect is that it prohibits all observation toward the celestial pole, whatever telescope it carries, owing to obstruction by the North support.

Photograph 1-4. *(upper left) German mount; b. (upper right) English mount; c. (lower left) Fork mount; and d. (lower right) Yoke mounting.*

I.4. — ELEMENTS OF OPTICS

I.4.1. — OBJECTS AT INFINITY

Consider a point source of light. From it rays of light emanate in all directions (Figure 1-9a).

Figure 1-9a.   **Figure 1-9b.**

Figure 1-9c.   **Figure 1-9d.**

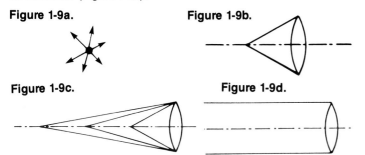

If we now observe this luminous point with a magnifying glass, we can direct our attention solely to the rays of light issuing from the object and penetrating the lens. These rays are all contained in a cone whose tip is the point source of light and whose base is the surface of the lens (Figure 1-9b).

If we increase the distance of the light source from the magnifying glass, we notice that the angle of the cone grows more and more acute (Figure 1-9c).

At the limit, if the observed object is sufficiently distant, we can consider our cone as practically a cylinder: all the rays of light reaching the lens are parallel to one another. Thus, we say that the object is "at infinity" (Figure 1-9d).

We say that an object (or an image) is at infinity when all the light rays we receive from each of its points can be considered parallel to one another.

In astronomy every object observed is considered to be at infinity. The nearest, the moon, is 380,000km distant, and the error caused by this approximation for the largest telescope in the world (6m in diameter) is only 3 thousandths of a second of arc!

I.4.2. — ANGULAR DISTANCE

The definition that we have just given means nothing more than that rays of light originating from two different points of an object at infinity should appear parallel to us. Thus, we do not see the North Pole and South Pole of the moon in the same direction: two rays of light each originating from one of these points together make an angle of ½°. So we say that the angular distance between the two lunar poles is ½° (Figure 1-10).

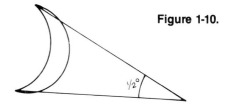

Figure 1-10.

We can consider the moon as being formed by a multitude of distinct points which we perceive in different directions, but all the light rays originating from the same point are actually parallel to one another. We say that the moon presents an apparent surface; this is also the case with the sun, all the planets except Pluto, galaxies, etc. The other heavenly bodies (stars, most planetary satellites) are considered as point sources.

I.4.3. — LENSES

The lenses usually found in astronomy or photography are divided into two categories:

Converging lenses, so called because they tend to concentrate the light they receive. These lenses share the characteristic of being thicker in the center than at the edge. There are three types of converging lenses, shown edge-on in Figure 1-11. Note that the converging lenses are represented in the optical sketches by a symbol shown on the far right of Figure 1-11.

Diverging lenses, which do just the opposite by dispersing the light. These lenses are thicker at the edge than at the center. Figure 1-12 shows the different types of diverging lenses and their symbolic representation:

In general, these lenses are pieces of glass whose surfaces are two spherical, coaxial disks (a plane can be considered as a sphere of infinite radius). The axis of symmetry for the lens is called the *principal axis.*

A lens is called "thin" when its thickness is small enough to be neglected in a geometric construction of the path followed by the light rays. Most of the time we shall assume that we are dealing with thin lenses. This approximation allows great simplification and does not change the results we shall be deriving in this book.

I.4.4. — THE CONSTRUCTION OF IMAGES

I.4.4.1. — THE OPTIC CENTER

The optic center of a thin lens is the point where it is crossed by its axis of symmetry. Neglecting the thickness of the lens allows us to present a very important law for the geometric construction of images: every ray of light passing through the optic center of a lens crosses it without being deflected (Figure 1-13).

I.4.4.2. — THE FOCUS AND FOCAL PLANES

Suppose that a point source at infinity lies along the principal axis of a converging lens it is shining upon. We therefore know that this lens receives a bundle of light rays parallel both to one another and to the principal axis. While passing through the lens, the light rays are deflected toward a point on this axis, which we shall call the "rear (*or* second) focal point" of the lens, where the point image of the object at infinity is thus formed (Figure 1-14).

If the object at infinity lies outside the principal axis, its image no longer forms on that axis, but on a plane perpendicular to it: the rear (*or* second) focal plane of the lens. The focal point is the particular point where this plane meets the principal axis (Figure 1-15).

Figure 1-11.

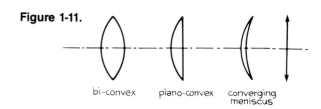

bi-convex   plano-convex   converging meniscus

Figure 1-12.

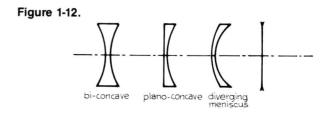

bi-concave   plano-concave   diverging meniscus

Figure 1-13.

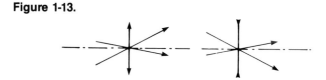

Figure 1-14.   **Figure 1-15.**

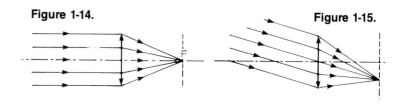

Conversely, we define a front (or first) focal point and a front (or first) focal plane: let there be a point source of light lying in this plane; after traversing the lens, the light rays emitted from this source are all parallel to one another (Figure 1-16).

Let's replace the converging lens in the preceding example with a diverging lens (Figure 1-17a). The parallel light rays striking it are no longer concentrated, but just the opposite—they are dispersed; however, they all seem to originate from a point to the left of the lens. For an optical system situated behind the diverging lens, the rays pass as if all the light the system receives were emitted from this point (Figure 1-17b). The diverging lens, therefore, has created at this point an image of the object at infinity, but since the light rays do not issue from this image, this image is qualified as virtual (as opposed to real). It would be impossible to make a virtual image appear on a screen as we have done for the real image of the magnifying lens experiment described in Section 1.2. The plane in which all the virtual images of objects at infinity are formed by a diverging lens is also known as a focal plane.

Likewise we can define a focal plane for the other side of the diverging lens, but the "objects" associated with it are themselves also virtual (Figure 1-18). They are no longer points which emit light, but places where a bundle of light rays would converge if it did not travel through the diverging lens.

The principles of foci and focal planes are not limited to lenses— they are valid for any optical system (Figure 1-19).

I.4.4.3. — THE GEOMETRIC CONSTRUCTION OF IMAGES

Knowing the rule of the optic center and the definition of focus, we have enough to easily construct the path of light rays through a lens. Here are a few examples.

Point object at infinity (Figure 1-20). We know that the image of this object is found in the focal plane. All the light rays emitted from the object converge toward this image—in particular, the ray passing through the optic center O. Now, we know that this ray is not bent; therefore, the image obtained can lie only at I (the intersection of the focal plane with the ray passing through O).

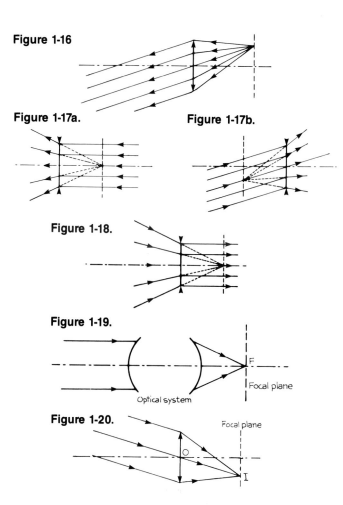

Figure 1-16

Figure 1-17a.

Figure 1-17b.

Figure 1-18.

Figure 1-19.

Optical system

F

Focal plane

Figure 1-20.

Focal plane

O

I

Nearby object (Figure 1-21). The image of the object AB created by the converging lens lies in a plane perpendicular to the principal axis. It is easy to find this plane by looking for the position of the point B', the image of point B. B' is determined from the intersection of the paths of two rays of light originating from B. The first, passing through O, is not bent. The second, which we shall choose as passing through F, is "seen" by the lens as originating from its front focus; hence, it re-emerges parallel to the principal axis. Thus is B' determined. The point A', the image of A, lies in the same plane as B' and, of course, on the principal axis.

Diverging lens (Figure 1-22). Let there be a virtual object AB lying between the diverging lens and its focal point F. We then construct the point B', image of the point B, by the intersection of the ray passing through O O and B, which is not bent, with the ray directed toward B and F, whose path becomes parallel to the principal axis after crossing through the lens.

I.4.4.4. — PRINCIPAL PLANES

In every optical system there exist two planes perpendicular to the principal axis, called the *first* and *second principal planes,* which are defined by the following property: if an object lies in the first principal plane, its image lies in the second principal plane, both object and image being the same height and on the same side of the axis of symmetry (the image is not reversed with respect to the object) (Figure 1-23).

This principle, though a little abstruse, is especially important for the formulas of Section I.4.4.7, where we shall need to measure the distance between the object and the principal planes.

In thin lenses, which concern us here, the two principal planes coincide at the optic center (Figure 1-24).

I.4.4.5. — THE PRINCIPLE OF MAGNIFICATION

Take an optical system (Figure 1-25) giving an image A'B' of an object AB:

By definition, the magnification μ of the system is the ratio of the height A'B' to the height AB of the object:

$$\mu = A'B'/AB,$$

which indeed corresponds to our intuitive notion of magnification.

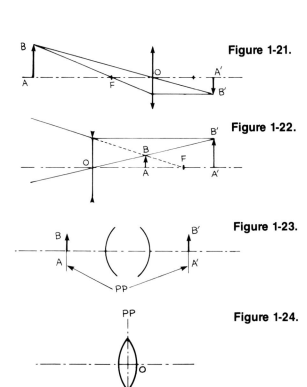

Figure 1-21.

Figure 1-22.

Figure 1-23.

Figure 1-24.

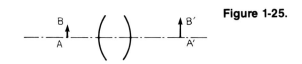

Figure 1-25.

By convention, the magnification is generally assigned a negative value when the image is reversed with respect to the object. This distinction has no importance in the framework of this book; in order to simplify the writing of formulas, we shall not make it. We shall consider all magnifications as positive numbers.

I.4.4.6. — FOCAL LENGTH

The first (or front) focal length is the distance separating the front focus from the first principal plane (Figure 1-26). Likewise, the second (or rear) focal length is the distance separating the rear focus from the second principal plane.

Figure 1-26.

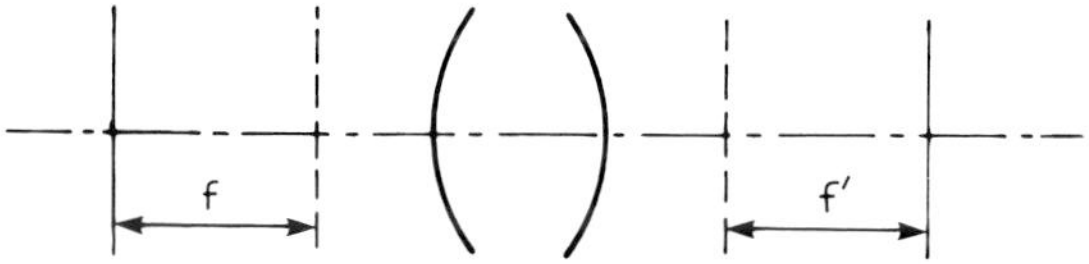

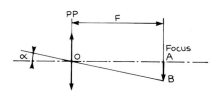

In general, these two quantities are identical and we speak of the system's *focal length.*

The focal length of a system is a very important characteristic; we shall come upon it frequently. In particular, it figures directly into the size of the images obtained.

The most important case for us occurs with an object at infinity, so that the image obtained in the rear focal plane has the value "a" such that:

$$a = F \bullet \alpha \ (*)$$

where F is the focal length of the optical system and α the angular diameter of the object expressed in radians.

Note that this formula is true only if α is a small angle (less than a few degrees) which will always be the case for us.

The value of an angle α in radians (rad) is equal to the length of an arc of unit radius and angle α (Figure 1-27).

Figure 1-27.

Since here the principal planes coincide with the lens, the focal length is the distance separating the optic center from the focus. Since the rays passing through the optic center are not bent, the triangle OAB with right angle A has angle O equal to α . Hence, AB = F • tan α . Since α is a small angle, we can write:

$$AB = F \bullet \alpha \ (\alpha \text{ expressed in radians}).$$

---

* Knowing that the circumference of an entire circle (that is, 360°) of radius 1 is 2 π , it is easy to convert from degrees to radians and vice versa:

(rad) = (2 π)/360 $\times \alpha$ (°);

(°) = 360/(2 π) $\times \alpha$ (rad).

It is easy to prove this formula for a thin lens (Figure 1-28).

Figure 1-28.

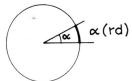

I.4.4.7. — SOME LAWS OF GEOMETRIC OPTICS

The three laws which we are going to present here without proof will help us calculate the size of images in optical systems with several lenses. They refer to an object AB not at infinity and its image A'B' through any converging optical system S of focal length f (Figure 1-29).

Figure 1-29.

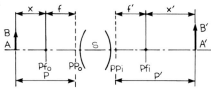

Now, p represents the distance from the object to the first principal plane and x the distance of the object from the front focal plane. Likewise, p' represents the distance of the image from the second principal plane and x' the distance of the image from the rear focal plane. All these quantities will be expressed in absolute value.

- Newton's law

    $$x \cdot x' = f^2$$

    If the object is at infinity, x is of course infinite. In order for the product x • x' to maintain a finite value, x' must be zero, which means that the image A'B' lies within the focal plane.

- law of conjugate relation of the principal planes

    $$1/p + 1/p' = 1/f$$

    If the object is at infinity, p is infinite and 1/p therefore zero. The conjugate relation is then written 1/p = 1/f or p' = f—again we verify that the image lies at the focus.

- expression of magnification

    We have seen the definition of magnification:

    $$\mu = A'B'/AB.$$

We can show that whatever the optical system considered, μ can be expressed as a relation of the distances to the principal planes by:

$$\mu = p'/p.$$

Once again it is easy to prove this formula in the simple case of a thin lens (Figure 1-30). Since the rays passing through the optic center O are not bent, we deduce that the triangles OAB and OA'B' are similar, hence the relation A'B'/AB = p'/p.

Figure 1-30.

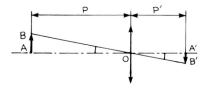

We can apply the relation $\mu = p'/p$ to the conjugate law we have just seen. In fact, $1/p + 1/p' = 1/f$ can be rewritten by multiplying the two sides of the equation by p':

$$p'/p + 1 = p'/f, \text{ or } \mu + 1 = p'/f, \text{ which we shall write}$$

$$\mu = p'/f - 1.$$

We shall have occasion to use this formula. It will help us calculate the magnification of a system of known focal length f when the distance p is not easy to measure.

Newton's law, the conjugate relation, and the expression of magnification apply as well to diverging systems. In order not to complicate notation, however, we have decided not to orient the principal axis. Thus, the letters p, p', and f represent distances always considered as positive. In the case of a diverging system, then, we must change a few signs in the preceding formulas: the formula for the conjugate relation is written $1/p - 1/p' = 1/f$ and the formula for magnification $\mu = p'/f + 1$.

We conclude these somewhat mathematical considerations by describing the change in the image given by a simple converging lens when we vary its distance from the object. The results, which we shall do no more than mention, are a direct consequence of the three laws we have just expressed (Figure 1-31).

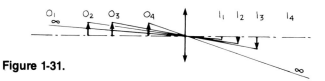

Figure 1-31.

When the object is at infinity (position O_1) there corresponds to it an image I_1 in the focal plane of the lens. If we bring the object nearer (position O_2) the image I_2 withdraws and at the same time grows larger. In O_3 and I_3 the object and image are both two focal lengths distant from the lens and are the same size. When the object passes from O_3 to O_4 (front focal point) the image withdraws toward infinity (I_4) and continues to enlarge.

I.4.5 — THE TELESCOPE

I.4.5.1. — MAGNIFYING POWER

Let us construct the image which a telescope gives an object at infinity with angle α (Figure 1-32).

Figure 1-32.

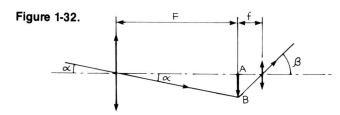

The objective of a telescope with focal length F forms an image of the object in its focal plane; this image has height $AB = F \cdot \alpha$. If we make the focus of the eyepiece coincide with that of the objective, then with the image AB the rays, emitted from A are all parallel to one another after crossing through the eyepiece. The same thing happens for those emitted from B. The eye behind the eyepiece thus has the impression of observing an object at infinity. But this object has for the eye an apparent angle β determined by the directions in which it sees A and B. The simplest way to know these directions consists of taking a ray emitted from each of these points and passing through the optic center of the eyepiece. We see that $AB = f \cdot \beta$, f being the focal length of the eyepiece.

The magnifying power (or angular magnification) M of the telescope (not to be confused with the *linear* magnification of the previous section! All you have to remember is that the notion of *angular* magnification applies to an object at infinity, whereas the notion of *linear* magnification is defined only for a nearby object, the first having to do with the angle of the object and the second with its actual size.) is defined most logically as being the ratio between the angle of the object seen through the telescope to the angle of the object seen without an instrument:

$$M = \beta / 2.$$

From the relations $AB = F \cdot \alpha$ and $AB = f \cdot \beta$ we get $F \cdot \alpha = f \cdot \beta$, that is $\beta/2 = F/f$, hence:

$$M = F/f.$$

The magnifying power of a telescope is equal to the ratio of the focal length of the objective to the focal length of the eyepiece.

I.4.5.2. — THE EXIT PUPIL

If the relation for magnifying power is generally well known by amateur astronomers, the notion of the exit pupil is not, which is regrettable because it occupies a fundamental place in choosing the magnifying power to use for visual observation. Since its importance is less evident in photography, however, we shall dwell only on its importance to visual observations.

Suppose that we are observing a star on the principal axis of the telescope. Let us trace the path followed by the rays emitted from this star which enter the telescope at the edge of its objective. We thus determine the envelope containing all the other rays gathered by the telescope (Figure 1-33).

Figure 1-33.

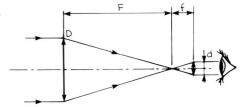

After passing through the focus, the outermost rays are bent by the eyepiece, from which they re-emerge parallel. In our drawing these outermost rays represent the boundaries of a cylinder (before crossing the objective) then of two cones (between the objective and the eyepiece) then another cylinder (after crossing the eyepiece)—inside which is found all the light originating from the star and gathered by the instrument. For the eye of the observer, the diameter d represents the maximum separation between two light rays leaving the telescope.

The relations between similar triangles permit us to write:

$d/f = D/F$ or $D/d = F/f = M$; hence,

$d = D/M$.

The exit pupil is equal to the ratio of the diameter of the objective to the magnifying power used.

Let us explain briefly why the notion of exit pupil is fundamental to visual observation:

For deep-sky observation, the smaller the magnifying power, the brighter the image, but the lower limit is the magnifying power M_e for which the exit pupil of the telescope equals the entrance pupil of the eye, or 6mm:

$M_e = D(mm)/6$.

Beyond this point, d is greater than 6mm and not all the light gathered by the telescope can enter the eye, which is wasteful. Further, as one grows older even this value becomes smaller.

For high-resolution observation (i.e. planets, double stars, etc.), the best results occur when the eye perceives a beam of light limited to diameter of 0.7 to 1mm. By substituting these values for d, we deduce that the optimum magnifying power for this type of observation is

$M = D$ to $M = 1.5D$,

where the diameter D of the objective is expressed in millimeters.

In photography you have to make sure that the diameter of each optical piece is greater than the exit pupil of the system ahead of it. This is the case, for instance, for the camera lens which would be placed behind the eyepiece of the telescope.

I.4.6. — MIRRORS

The principle of the telescope optics teaches us that a mirror can play the same role as a lens. A concave mirror is convergent; a convex mirror is divergent. Apart from the plane diagonal mirror of the Newton telescope, whose role is simply to reflect a beam of light 90°, we find two types of mirrors in astronomy:

• *parabolic concave mirrors* (Figure 1-34) give a point image at their focus of a point object at infinity. They are the objective of reflecting telescopes. In place of a parabolic mirror, certain types of telescopes use a spherical concave mirror with similar characteristics.

• *hyperbolic convex mirrors* (Figure 1-35) have an enlarging effect on the image of Cassegrain telescopes. This diverging mirror must be a hyperbolic section in order to give a point image of a point object at a finite distance.

These mirrors are characterized by an optic center, principal planes, focal planes, focal lengths, etc. We can apply to them the relations that we have seen for lenses, as well as the same geometric constructions, with one exception: the light does not pass through the optic system, but is reflected in the direction from which it comes.

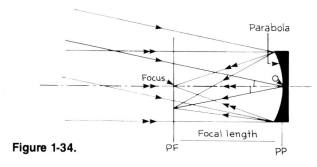

Figure 1-34.

Parabola

Focus

Focal length

PF   PP

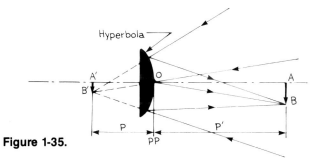

Figure 1-35.

Hyperbola

I.4.7. — DIFFRACTION

The laws of wave optics show that every time a beam of light crosses an aperture, it undergoes a slight dispersion (Figure 1-36).

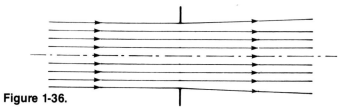

Figure 1-36.

Since the objective of a telescope has a limited diameter, it also behaves like an aperture; consequently, even if the optics are perfect, all the parallel light rays collected are not exactly concentrated into one point, but into a spot of small dimension called a *diffraction disk*.

The image of a star does not appear as a point, but as a small disk—where the greatest part of the light is concentrated, surrounded by rings a little less bright (Figure 1-37a,b)

Figure 1-37a.

Figure 1-37b.

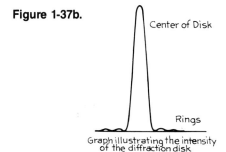

Center of Disk

Rings

Graph illustrating the intensity of the diffraction disk

It is this diffraction pattern which limits the theoretical resolving power, since we will never be able to dissociate two planetary details or two stars whose separation is too small compared to the width of the central disk. Now, in angular value the width of the central disk is inversely proportional to the diameter of the objective (the larger the aperture, the less it scatters the light). This explains the expression 120/D that we gave in Section I.3.3 for the value of the theoretical resolving power of a telescope of diameter D in millimeters. In fact, the radius of the central disk is 140", a value slightly greater than the resolving power. Practice has shown that we can "separate" two stars whose disks partially overlap.

Figure 1-37c.

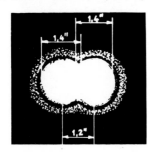

Figure 1-38 shows a diffraction disk. It is not the image of a star in a telescope—which otherwise would be similar, but too small to be reproduced here—but a figure obtained in the laboratory by a laser point source passing through a small-aperture (0.3mm).

Figure 1-38.

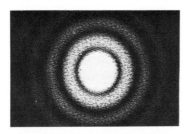

I.4.8. — CHROMATIC ABERRATION

I.4.8.1. — THE CAUSES OF CHROMATIC ABERRATION

All transparent bodies have an index of refraction whose value varies slightly according to the wavelength of the light passing through them. For the visible spectrum, the indices for glass and for air increase when the wavelength diminishes. Therefore, when blue rays enter one of these media, or when they emerge, they are bent more than the red rays.

This property allows a prism to dispense light by separating the rays of different colors (Figure 1-39).

Figure 1-39.

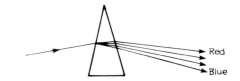

Another consequence is that every lens focuses rays of different wavelengths into slightly different points (Figure 1-40). It is therefore impossible to obtain a sharp image for all wavelengths at once.

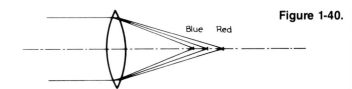

Figure 1-40.

I.4.8.2. — THE ACHROMATIC OBJECTIVE

By bringing together a greatly converging lens with a slightly diverging lens made from a different glass whose coefficient of chromatic dispersion is greater than than of the glass of the converging lens, we obtain a converging combination whose chromatic aberration has almost entirely disappeared. In fact, the chromatic aberrations of the two lenses compensate for each other. The objective thus created is called "achromatic."

Astronomical observation is impracticable with a single-lens objective. Depending on the focus setting, you obtain either a bluish image surrounded by a red halo or a reddish image surrounded by a blue halo.

Consequently, achromatic objectives are by far the most common; however, they do present a slight residual aberration which can be troublesome for high-resolution photography.

Three-lens objectives, called "apochromats," also exist. Their correction for chromatic aberration can be considered total for the visible spectrum. These objectives, however, are not very widespread because they are relatively expensive. Figure 1-41 shows the chromatic aberration relationships for the three objectives.

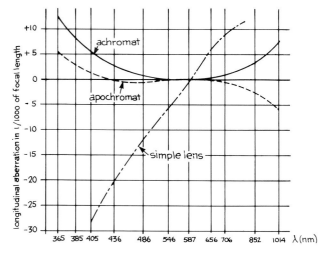

Figure 1-41.

I.4.8.3. — CHROMATIC ABERRATION OF THE ATMOSPHERE

Rays emanating from a star far from the observer's zenith strike the atmosphere obliquely. Thus, they undergo refraction and a slight chromatic dispersion, owing to the variation in the index of refraction for air with respect to wavelength. The attentive observer at the telescope will notice that planets near the horizon are surrounded by a blue halo on their higher portion and by a red halo on the lower (Figure 1-42).

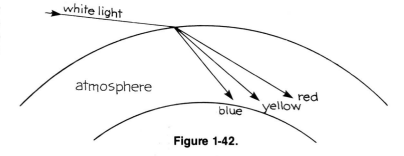

Figure 1-42.

This atmospheric aberration can alter high-resolution images for stars less than 40° to 50° from the horizon.

CHAPTER II

PHOTOGRAPHY WITHOUT A TELESCOPE

It is possible to take astrophotographs with a simple camera and a standard or telephoto lens? We shall see that we can answer this question affirmatively, so long as we forget about high resolution. And if when we say "photography without a telescope," it means that the image fixed on film has not been produced with the help a telescope can give—and the latter can prove to be a very effective aid. . . .

II.1. — IMAGE SIZE

As we have indicated in the review on geometric optics, the size "a" of the image of an object with angular diameter α produced by an objective of focal length F is

$$a = F \cdot \alpha ,$$

where α is expressed in radians, and a and F in the same unit, usually millimeters.

More concretely, we know that the moon (α = 0.01 rad, or 30°) photographed with a focal length F of 1000mm gives an image of diameter a $\cong$ 10mm. From this example we can quickly calculate the size of the image of any celestial object with any objective by using a simple "rule of three."

We find that the moon, photographed with a standard 100mm telephoto, will have a diameter of 1mm on the negative. As for Jupiter, whose angular diameter is one fortieth of the moon's, it will be represented by an image of less than 30 microns! Therefore, it is quite clear that it is impossible to photograph fine planetary detail with conventional photographic equipment. Even using a more powerful telephoto would not provide a satisfactory solution.

II.2. — STELLAR FIELDS

The more a given angular diameter is represented by a small area on the film, the more a given area of the film represents a large angular diameter. In other words, the shorter the focal length of the lens, the lower the resolution of the image, but the larger the field that is photographed.

For astronomy a standard camera must therefore be considered as a wide-field camera (5° to 60°): It is best suited to photographing "stellar fields," that is, sundry configurations showing stars in their relative positions.

It is interesting to photograph in this way various constellations, planetary conjunctions, extended nebulae, or galactic fields. Hence, for these you would choose a relatively wide field, using a 35 to 85mm lens (for 35mm film). The Milky Way region lends itself quite well to this type of photography because the stars are abundant and a photograph gives the impression of a spectacular swarm with an exposure of only a few minutes.

If you are interested in a more restricted region of the sky—a star cluster like the Pleiades, for example—then a 100 to 300mm telephoto lens becomes useful.

As you will see later when we discuss deep-sky photography, you might be induced to work with fine-grain, hypersensitized films (Kodak 2415). Thanks to hypersentizing, it is thus possible, by greatly enlarging the negative (30 or 40 times) to use a 200 or 300mm telephoto lens to obtain pictures of a quality comparable to those obtained at the focus of a 750--1200mm focal length telescope on grainier films.

Generally, the number of stars visible on the photograph after a long exposure will be greater than the number of stars visible to the naked eye. You can thus look for planets invisible or barely visible to the unaided eye, such as Uranus, Neptune, Ceres, Pallas, etc., and mark their location. You can just as easily make yourself a veritable sky map, relatively precise and detailed by joining together a series of these large-field photographs.

Photograph 2-1. On the opposite page is a photo showing the south of the constellation Centaurus; at the right is the Southern Cross and the Coal Sack. 50mm focal length objective; F/D = 2.8 exposure: 8 minutes. Taken by Mosser on Ile de la Reunion.

While making long-exposure photographs with a large field, a meteorite may cross part of the sky you are photographing. The meteor will appear on the film in the form of a streak that marks its trajectory against the starry background. Thus, it is possible to take spectacular pictures in the course of a meteor shower.

Finally, this type of photography allows you to hunt large comets when one of them does us the honor of crossing our sky.

II.3. — THE APPARENT MOTION OF THE SKY

Since the Earth rotates on its axis from West to East every 24 hours, we have the impression that the sky turns about us from East to West with the same speed.

A star at the celestial equator thus travels 360° in 24 hours, 15° in an hour, or 15' *in one minute. Hence, an angle of 30'*, equivalent to the moon's diameter, is traveled in two minutes. If we take a photograph of the starry sky with a 100mm objective mounted on a fixed support, after only 2 minutes exposure the stars will be displaced by 30' and will therefore no longer be represented on the film by points but by streaks 1mm long!

Notice that the apparent motion of the sky is less and less marked as you move from the celestial equator toward the pole, where it is nil. This is shown in Figure 2-1.

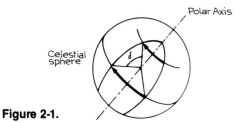

Figure 2-1.

If the exposure time is expressed in minutes, the apparent displacement at the equator has the value in minutes of arc of:

$d = 15 \times t$; and for any declination δ :
$d = 15 \times t \times \cos \delta$ (at the equator, $\cos \delta = 1$; at the pole, $\cos \delta = 0$).

The effect is not always unpleasant to let the stars "streak" by ridgidly mounting the camera during a long exposure. If the region photographed is the celestial pole, you can see all the stars draw arcs of circle around the polestar, which is practically immobile.

II.4. — THE PIGGYBACK MOUNTING

By contrast, the displacement of stars on the film is not acceptable for most of the tasks previously cited. The camera must therefore follow the apparent motion of the sky.

The simplest set-up consists of fixing our camera on a telescope equipped with an equatorial mounting, both aimed at the same region of the sky. This is what is called the "piggyback" mounting.

Photograph 2-2. *A camera mounted "piggyback" on an equatorial mounted refractor.*

Photograph 2-3. *Comet West (opposite page) showing a tail 3° long. Taken by J. Silvain on 3-14-76 with a 500mm telephoto using 103aE film, exposed at f/6.3 for 10 minutes.*

With the help of a telescope, you aim at a guide-star which you make coincide with the center of a reticle placed at the focus. During the entire exposure, the telescope (and thus the camera) turns about the mounting's polar axis so that the guide-star is kept well within the center of the reticle. If you do not use a motor, the turning can be done manually by slowly moving the right ascension of the mounting. The shorter the camera's focal length, the more imprecise the tracking can be. By way of example: with a 100mm objective, correcting every 5 seconds creates a blur 40 microns long on the film, which is practically imperceptible on ordinary film.

II.5. — OTHER POSSIBLE MOUNTINGS

For counteracting the rotation of the sky we have just mentioned the luxurious solution. It is the most effective, but it assumes that the photographer has a telescope.

In fact, we have simply to attach the camera to a platform able to turn about an axis parallel to the polar axis. The orientation of the camera on the platform, which determines the region of the sky sighted, can be done with the help of simple wooden blocks, the aim not requiring great precision. Every handyman can build the mounting, the principle of which is described in Figure 2-2.

The nut moves along the axle driven by the motor and pushes the arm. The arm moves the platform about the polar axis.

Another possible solution is the mounting shown in Figure 2-3, where the polar axis is simply made up of a row of hinges.

In both the mountings we have just presented, the motor does not permit precise tracking beyond a few minutes' exposure. You should add a small guide scope, moveable with the camera, in order to control possible drifting. Here you can also replace the motor with a manual action, in which case the guide scope becomes indispensable.

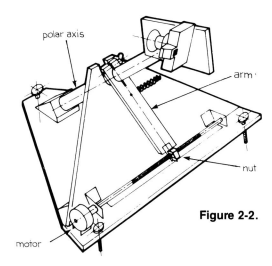

Figure 2-2.

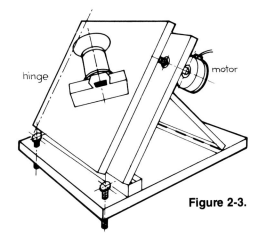

Figure 2-3.

II.6. — HOW TO PROCEED?

The choice of lens depends upon the field desired. With a 35mm camera, the gamut of focal lengths typically extends from 35mm to 135mm.

The Table 2-1 gives the field obtained with a 35mm camera and the corresponding focal length of the lens:

TABLE 2-1
Field of View for Various 35mm Lenses

Focal Length (mm)	Field
24	74° × 53°
28	65° × 46°
35	54° × 38°
50	40° × 27°
85	24° × 16°
100	20° × 14°
135	15° × 10°
200	10° × 7°
300	7° × 5°

In order to gather the most light you want to keep the lens as wide open as possible and not stop it down.

Preferably you should choose a relatively fast film, either black-and-white or color. Color films perceive the color differences among stars better than the eye. Color photographs of constellations are usually very beautiful.

Typical exposure times range from a few minutes to a few dozen minutes. The exposure duration depends upon the film's speed, the desired limiting magnitude, sky conditions, and . . . the patience of the operator.

Photograph 2-4 a and b. The comet Iras-Araki-Alcock taken by H. Le Tallec on 5-11-83 with a 135mm telephoto using 103aF film exposed at f/2.8 for 10 minutes. The camera was piggy-back mounted on a telescope so that the comet could be tracked since the comet's motion is independent from the stars — note that while the comet is sharp the stars are represented by trails. Comparing these two shots, taken 15 minutes apart, reveals the comet's displacement with respect to the stars in the field.

II.7. — LIMITING MAGNITUDE

An important given in this type of photography is the number of stars that you will be able to record on film, or (exactly the same problem) the luminosity of the faintest star that you can hope to capture: this is what we call *limiting magnitude*. Magnitude is the value characterizing the brightness of a star; the higher the value, the fainter the star. The scale is such that a star of magnitude m is 2.5 times brighter than a star of magnitude m + 1. The brightest stars are of magnitude 0 to 1; the faintest stars visible to the naked eye approach 6; as for a planet like Venus, it can attain a magnitude of − 4.3 at greatest brilliancy. The limiting magnitude you can photographically record therefore depends greatly upon the film lens, and the length of exposure.

We shall demonstrate that for objects presenting an apparent surface (Section IV.7) there is a relationship between the object's luminosity to the diameter and focal length of the objective, the sensitivity of the film, and to the length of exposure. But this relationship involves the size of the image on the film, that is, the product of the focal length and the angular diameter.

Now, since stars are point sources, the angular diameter is zero, so that the size of a stellar image on film depends not on the focal length, but on the diffraction image (related to the diameter of the objective), the quality of the optics, and the response of the film to almost pin-point lighting. The law relating extended objects, therefore, no longer applies and we have to take into account other complex phenomena.

In order to calculate the limiting magnitude, amateur astronomers tend to use the following empirical law:

$$M = 8.4 + 5 \log D + 2 \log T - \log F + 2.5 \log (S/800),$$

where M is the limiting magnitude, T the exposure in minutes, D and F the diameter and the focal length of the objective in centimeters and S the speed of the film in ASA. In the literature, we have always seen this law written without the last term; thus, it is held to be valid for a film of ASA 800. We have seen fit to add the term 2.5 log(S/800), which permits us to make the calculation for all films, provided that they show nearly the same reciprocity failure (see Section V.3.1) and have medium grain and contrast. Thus, this law applies essentially to ordinary films. The use of special films, whose response is appreciably different, will be covered in the chapter devoted to faint-object photography.

A discussion of this empirical law is given in Appendix A. You should be well aware that the calculation of limiting magnitude has only an indicative value—the law presented can not take into account every variable, and the error can easily reach an entire magnitude. Nevertheless, the law is useful for knowing a priori what you can expect from a photograph Table 2-2 will give you a quantitative idea of the possibilities of each lens without having to make all the calculations to see the limiting magnitude

TABLE 2-2
Limiting Magnitude for Various F-Stop Focal Length Combinations for 1 Minute Exposures on ASA-800 Film

f-stop	Focal Length	24	35	50	85	100	135	200	300	400	500
1.4		9.2	9.8	10.5	11.4						
2		8.4	9.1	9.7	10.6	10.9	11.4				
2.8		7.7	8.3	9.0	9.9	10.2	10.7	11.4			
4		6.9	7.6	8.2	9.1	9.4	9.9	10.6	11.3		
5.6								9.9	10.6	11.1	11.5
8									9.8	10.3	10.7

for a 1 minute exposure on ASA 800 film cut various f-stops and focal lengths F:

If the film is not developed at ASA 800 or if the exposure is not for 1 minute, you will need to add the corrective values shown in Tables 2-3 and 2-4:

TABLE 2-3
Correction Factors for Film Developed Other Than ASA800

S (ASA)	M
50	− 3.0
100	− 2.3
200	− 1.5
400	− 0.8
800	0
1600	+ 0.8
3200	+ 1.5

TABLE 2-4
Correction Factors for Film Exposed Other Than 1 Minute

T	M
10 s	− 1.6
30 s	− 0.6
1 mm	0
2 mm	+ 0.6
5 mm	+ 1.4
10 mm	+ 2.0
20 mm	+ 2.6

Photograph 2-5. The clusters h and chi Persei taken by J.P. Trachier with a 135mm telephoto on Tri X film exposed for 40 minutes.

You can not prolong the exposure indefinitely, however. In fact, even an apparently black sky gives a certain residual luminosity which winds up creating a slight fog on the film—this fog obscures the faint stars. Thus, too long an exposure is self defeating when one wants to record faint stars. Therefore, the maximum exposure for a given film (here ASA 800) depends upon the ratio F/D (f-stop). Empirically we can derive the law

$$^{T}max = 1.5 \, (F/D)^{2.5} \, ,$$

where the time is expressed in minutes. Notice that when dealing with sky-fogging, the law of luminosity of extended bodies applies (see the discussion in Appendix A) and should give a maximum exposure proportional to $(F/D)^2$. The power 2.5 instead of 2 simply takes into account reciprocity failure (see Section IV.3.1).

In the author's opinion, it is also necessary to account for the speed S of the film, expressed in ASA, in the following manner:

$$^{T}max = 1.5 \, (F/D)^{2.5} \, (800/S)^{1.25}.$$

Table 2-5 gives an indication of maximum exposure times according to the film speed and the f-stop used. We emphasize that it is only an indication; indeed, if the sky is slightly hazy, or if the moon is present, the film will fog much more rapidly (on this subject you should remember that it is practically impossible to take good stellar pictures during periods of the full moon).

TABLE 2-5
Indication of Maximum Exposure Time for Given Film Speed and F-Stop

Film Speed	f-stop	1.4		2		2.8		4		5.6		8	
		hr	min	hr	min	hr	min	hr	min	hr	min	hr	min
50		1	50	4	30	10	30						
100			45	1	55	4	25	10	45				
200			20		50	1	50	4	30	10	30		
400			8		20		45	1	55	4	25	10	45
800			3		8		20		50	1	50	4	30
1600			1.5		4		8		20		45		55
3200			.5		1.5		3		8		20		50

Now let's calculate the attained magnitude for each type of lens when using the maximum exposure warranted by the sky-fog. Entering the expression for this maximum time in the equation for the limiting magnitude, we obtain:

$$^{M}T_{max} = 8.75 + 4\log F \quad (F \text{ in centimeters}).$$

The attained magnitude when using the maximum exposure called for depends neither on the objective diameter nor on the film speed, but only on the focal length used. Surprising at first glance, this result is explained by the fact that only the contrast of the star against the fog makes any difference. Now, the objective diameter and the speed of the film influence the acquisition of fog and starlight in the same way and thus play no part in the contrast. On the other hand, the focal length causes a different dispersion of light for a point-source star than for a surface luminosity. Table 2-6 gives the attained magnitude for a maximal exposure for various common focal lengths.

TABLE 2-6
Attainable Magnitude for Maximum Exposure Using Common Focal Lengths

Focal Length	24	35	50	85	100	135	200	300	400	500
Maximum Magnitude	10.3	10.9	11.5	12.5	12.8	13.3	14.0	14.7	15.2	15.5

Up to now all the magnitudes calculated are valid for photography near the zenith. But it must be remembered that the Earth's atmosphere absorbs significant amount of light. This absorption becomes more appreciable as the thickness of the air crossed by the light increases, and this thickness increases the farther we aim from the zenith (Figure 2-4).

Photograph 2-6. On the opposite page is the Milky Way in the constellation Aquila taken by J. Silvain with a 50mm lens, F/D = 2 on 103aO film exposed for 8 minutes.

Figure 2-4.

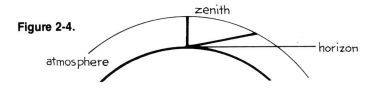

Therefore, we must take into account the increasing loss in magnitude the farther we go from the zenith. The Figure 2-5 gives the value of the drop in attained magnitude, with respect to the distance from the zenith, for a normal atmosphere, excluding any haze or urban pollution.

Figure 2-5.

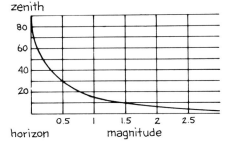

II.8. — PHOTOGRAPHING METEORS

Meteor hunting is one particular application of wide-angle stellar photography. It consists of photographing a large region of the sky and . . . waiting for a "shooting star" to pass through the field and leave a luminous trail on the photograph. The goal can be aesthetic, but such a picture can serve to determine the trajectory of the meteorite (several photographs taken from different locations are required). If the camera is fixed, the stars leave trails and the photograph reveals the trajectory of the meteorite in relation to the sun; if the camera is in motion, the film records the displacement of the meteor in relation to the constellations.

Photograph 2-7. *This double flare-up of a Perseid meteor was taken by Paul Moithy using Tri X film with the camera mounted on a fixed tripod.*

Photograph 2-8. *This photograph shows a meteor of magnitude of −3. It was taken by Christian Buil with a 50mm f/1.7 lens. The background star field shows that the camera was guided during this exposure.*

The angular velocity of the meteorite can be determined with the use of a notched disk turning at great speed in front of the lens (Figure 2-6).

Figure 2-6.

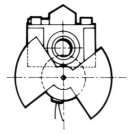

Photograph 2-9. *This photograph shows a meteor trail which has been occulted by a rotating notched disk. It was taken by Christian Buil using a 35mm f/2.8 lens. The period between two successive occultations is 0.06 second.*

The notched disk occults the objective at a known frequency, for example, 10 times per second. Thus, the trail of the meteorite appears as a dotted line, each segment corresponding to the time during which an opening passes in front of the lens. In this way we can calculate the angular velocity of the meteorite and the duration of its combustion.

Meteor hunting is the most fruitful during periods of "meteor showers." The principal streams are shown in Table 2-7:

- Quadrantids: January 1-5
- Lyrids: April 20-22
- Aquarids (γ): May 1-13
- Aquarids (δ): July 25-August 10
- Perseids: July 30-August 19
- Draconids: October 8-11
- Orionids: October 16-22
- Leonids: November 15-20
- Geminids: December 9-18

It is advisable to use a wide-angle lens (focal length less than 50mm). Do not aim at the radiant of the shower (the point from which all the meteors seem to originate — Figure 2-7 — but where the trails are not yet visible) but at a zone displayed from the radiant by 45° to 90° (Figure 2-8).

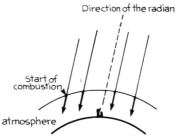

Figure 2-7. *Meteorites from the same shower have nearly parallel trajectories.*

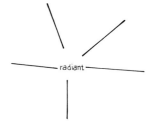

Figure 2-8. *Because of perspective, the observer sees radial light streaks, but no phenomena in the direction of the radiant.*

CHAPTER III

PHOTOGRAPHY WITH A TELESCOPE: OPTICAL SYSTEMS

We have just seen in Section II.1 that the resolution necessary for most objects studied in astronomy leads us to look for long focal lengths. At this level, the telephoto lenses of photographers become insufficient, but we shall see how telescopes can furnish us with the long focal lengths we require.

Here we are, then, at the heart of the subject, and the problems to resolve are going to be more complex than those mentioned for photography without a telescope. Also, inasmuch as we have treated the latter subject in a relatively complete fashion in Chapter II, the various aspects of using a telescope for astronomical photography and the problems associated with it will occupy the rest of this book.

III.1. — PRIME-FOCUS PHOTOGRAPHY

We know that the objective of a telescope forms in its focal plane an image of the observed object. All we have to do to capture this image then is to place photographic film in this focal plane. In order to build this system, we mount the body of a camera (without its lens) so that the film plane is at the telescope's prime focus—hence the name "prime-focus photography" (Figure 3-1).

Telescopes (refractors or reflectors) used by amateurs commonly have focal lengths from 1 to 2 meters, which represents an appreciable gain over typical telephoto lenses. There are telephotos with focal lengths of

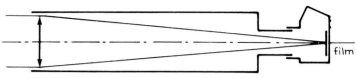

Figure 3-1.

1 or even 2 meters, but their cost is much higher than that of a telescope of equal performance. Furthermore, for such focal lengths the problem of the apparent rotation of the sky can no longer be treated as simply as in the case of photography without a telescope—now the use of a good equatorial mounting is necessary.

In many cases the size of the images obtained at the focus of a telescope is still insufficient. For example, Jupiter's image, would be only 0.3mm long with a 1m focal length. As a result, we have to use enlarging systems which allow us to increase the image size furnished by our telescope.

III.2. — THE PRINCIPLE OF EFFECTIVE FOCAL LENGTH

We have already expressed the relation of the size "a" of an image to the angular diameter α of the object and to the focal length F of the optic used:

$$a = F \times \alpha$$

Notice that at this level we are interested in the focal length only because it represents the relation between the dimensions of the image and the object.

Let's take the case of any optical system for which we do not know all the parameters and observe an object of angle α. The only thing that concerns us a priori is to know the size "a" of the image which the optical system furnishes us. We can thus say that everything happens as if the image of the object of angle α had been obtained with a simple objective of focal length F_e such that:

$$a = F_e \times \alpha$$

The focal length F_e is called the "equivalent focal length" or the "effective focal length" of our optical system—and for the time being this is the sole characteristic we need.

Of course, in the case of prime-focus photography, the effective focal length of the system is by definition equal to the focal length of the telescope used.

We can now make the setup shown in Figure 3-2. A telescope gives us at its focus an image AB of an object of angle α. If F_o is the focal length of the telescope, then $AB = F_o \times \alpha$. We place behind this telescope any optical system S which creates another image A'B' of the observed object from the primary image AB. In the general case, A'B' will be larger than AB, and the ratio $\mu = A'B'/AB$ is called the magnification of the system S.

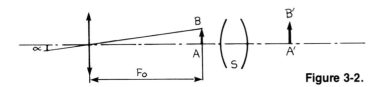

Figure 3-2.

A'B', then, is the image of the object of angle α given by the combination of telescope plus enlarging system. The effective focal length of this assembly is thus the value F_e such that $A'B' = F_e \times \alpha$. Now, A'B' $= \mu \cdot AB$ and $AB = F_o \times \alpha$, hence:

$F_e = F_o \times \mu$.

The effective focal length of the system is equal to the prime focal length of the telescope multiplied by the magnification of the enlarging system used.

We are now going to describe some optical systems which allow us to enlarge the primary image furnished by the telescope—that is, to utilize an effective focal length greater than the nominal focal length.

III.3. — A NEGATIVE ENLARGER: THE BARLOW LENS

The Barlow lens is a diverging lens placed in the light path just before the focus of the telescope. Good-quality Barlow lenses are usually not constructed with a single lens, but with two lenses to attenuate chromatic aberration.

In treatises on optics, the focal length of diverging lenses is conventionally noted with a negative value, hence the name "negative enlarger" for the Barlow.

Figure 3-3 shows that the Barlow lens "lengthens" the path of the light rays and diminishes the angle of the cone of light arriving at the new focus. Let's see how we can calculate the magnification of a Barlow.

Figure 3-3

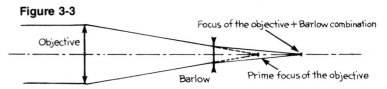

In Figure 3-4 object AB is the primary image given by the telescope; thus, the focus of the latter is found at A. Let F be the focus of the Barlow (thus OF = f, the focal length of the Barlow). The ray (1) is parallel to the principal axis after crossing through the lens, since it was converging toward F. The ray (2) passes through the optic center O; thus, it is not bent. Rays (1) and (2), which were converging toward B before crossing through the lens, will intersect at B', the image of B. This is sufficient to determine the position of the image A'B' of AB. A few simple geometric considerations gives us the position of A'B', knowing the relative positions of the prime focus and the Barlow's focus and focal length:

$1/OA' = 1/OA - 1/f$

(we see again the conjugate relation formula of diverging systems, $1/p - 1/p' = 1/f$, from Section I.4.4.7).

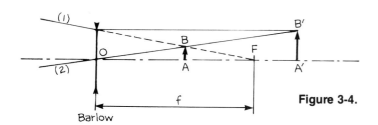

Figure 3-4.

The magnification obtained is by definition the ratio A'B'/AB or OA'/OA, or $\mu = F/(f-OA)$ or $\mu = 1 + OA'/f$.

We see that we can change the magnification by varying the distance OA. The focus of the telescope (object AB) must always lie between the Barlow lens and the focus of the Barlow, but the closer together the two foci are, the greater the magnification is. If the two foci are identical (A and F coincide) the image A'B' is projected to infinity: we can no longer photograph it, but it is fit for visual observation. This is the principle of opera glasses and . . . the Galilean telescope.

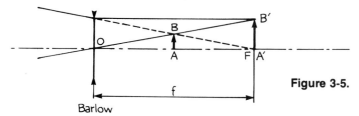

Figure 3-5.

Barlow lenses, however, are designed to furnish the best image at a specific magnification. For the majority of commercial lenses, the optimum corresponds to the case when the amplified image is found at the focus F (F and A' identical). So we see that the telescope focus A must be found exactly in the middle of the segment OF—thus, the magnification is 2 (Figure 3-5). These lenses generally carry the mention "2 ×" on the barrel. You can also find Barlow lenses that produce a magnification of 3.

III.4. — AFOCAL METHOD: AMPLIFICATION WITH EYEPIECE AND CAMERA LENS

This method is the most intuitive way of photographing with a telescope. Let's begin by considering a telescope used in standard fashion for visual observation (Figure 3-6).

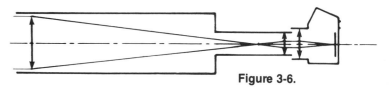

Figure 3-6.

From a star (point object at infinity) the telescope receives parallel rays of light. The rays of light leaving the eyepiece are also parallel, since the eyepiece is adjusted so that its front focus coincides with the rear focus of the objective. Thus, the eye always sees the star at infinity, but enlarged by a factor $M = F_o/f$, where F_o and f are the focal lengths of the objective and the eyepiece.

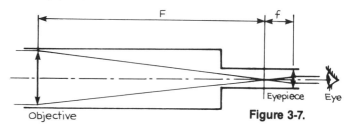

Figure 3-7.

It seems completely natural to replace the eye with a camera lens and camera (Figure 3-7). If the camera lens is set at infinity, it focuses on the film the rays of light emitted from the eyepiece. In this way an image of the observed object is obtained.

If this object presents an angle α, it is perceived behind the telescope with an angle $M \times \alpha$, where M is the magnifying power of the telescope. The camera thus "sees" an object of angle $M \times \alpha$. So the camera lens, with focal length F_c, gives an image of size a such that a = $F_c \times (M \times \alpha)$. The effective focal length F_e of the system is therefore $F_c \times M$.

Now, $M = F_o/f$, hence $F_e = (F_c \times F_o)/f$.

F_o being the prime focal length of the telescope, the magnification of the eyepiece-lens system is thus

$$\mu = F_c/f.$$

The magnification is equal to the ratio of the focal length of the camera lens to that of the eyepiece.

For instance, if we use a telescope of focal length 1200mm with an eyepiece of 20mm and a 50mm lens, we have a magnification:

μ = 50/20 = 2.5, and an effective focal length
F_e = 1200 × 2.5 = 3000mm.

Instead of a telescope, a similar arrangement can be made by placing binoculars in front of the camera lens. For instance, 12 × binoculars coupled with a 50mm lens gives an effective focal length of 12 × 50 = 600mm.

III.5. — POSITIVE PROJECTION: AMPLIFICATION WITH EYEPIECE ALONE

If the image observed by the eyepiece is not formed at its front focus, but slightly ahead (that is, if the eyepiece is pulled back slightly from the position where its front focus coincides with the rear focus of the telescope objective) the light rays which have passed through the eyepiece are no longer parallel, but converge to form a new image (Figure 3-8).

Figure 3-8.

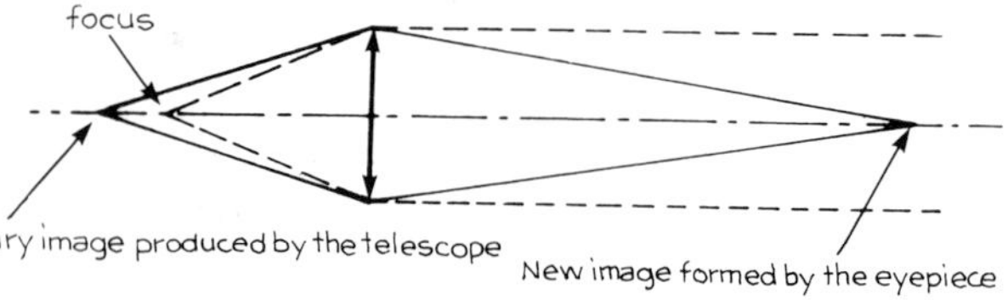

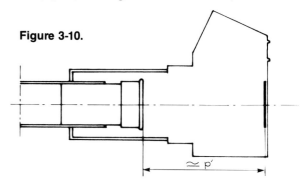

focus

Primary image produced by the telescope   New image formed by the eyepiece

It is thus useless to employ a camera lens—the eyepiece alone is enough to create an enlarged image from the telescope's primary. Figure 3-9 shows this setup.

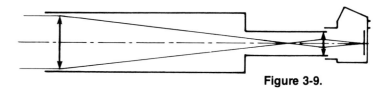

Figure 3-9.

We saw in Section I.4.4.7 that the magnification u provided by this assembly is

$$u = p'/f - 1,$$

where f is the focal length of the eyepiece and p' the distance between the obtained image and the image plane of the eyepiece. In general, this image plane is close to the eye lens of the eyepiece and we can prove that in every case the error caused is less than the focal length of the eye lens of the eyepiece. Moreover, when the focusing takes place, the plane where the image forms must coincide with the film. Thus p' represents the distance separating the film from the eye lens. For example, if this distance is 180mm, with a 20mm eyepiece the magnification obtained is μ = 180/120 − 1 = 8.

Figure 3-10.

$\simeq$ p'

III.6. — OTHER ENLARGING SYSTEMS

The world of terrestrial photography offers us equipment capable of enlarging the primary image of the objective.

First, are teleconverters, which, mounted between a camera lens and body, multiply the focal length of the lens two or three times. These optics, comparable to Barlow lenses, are not recommended because they generally are of lower quality. A teleconverter is designed to have an even quality over an extended field. However, in astronomy we require excellent quality over a very narrow field. Barlow lenses are intended for this role and are therefore better adapted to astronomical problems.

The major manufacturers of photographic equipment (Nikon, Canon, Olympus, etc.) provide special lenses for microphotography. They are relatively small optical systems mounted on the camera body with bellows to allow for focusing. They are designed to give an image 2 to 10 times larger than the object; thus, at this level of enlargement, they can efficiently replace systems using an eyepiece, since they are themselves calculated to maintain image quality. They have two shortcomings, however: a very high price and the necessity of a heavy housing with a bellows rail, which an amateur telescope will have trouble supporting.

III.7. — FOCAL REDUCERS

As the name indicates, this is a system which does just the opposite of the enlarging systems we have just seen, since its role is to furnish an effective focal length shorter than the prime focal length of the telescope. Opticians say that the magnification of such a system is less than 1 (image smaller than the object).

A focal reducer (or "telecompressor") is sometimes used to photograph extended bodies of faint luminosity. They thus permit us to concentrate more light on the same surface of film. The drawback is a shorter focal length, that is, a loss of resolution with prime-focus photography. In addition, inasmuch as the field covered by the film is larger than the decrease in focal length, a focal reducer generally reduces the portion of the field with good optic quality.

Reducers can be used with a converging lens. In general, they provide a magnification of 0.5: the image is twice as small after passing through the reducer.

Two arrangements are possible for obtaining this result: you can put the converging lens either after the prime focus like an enlarging eyepiece or in front like a Barlow lens. Figure 3-11 shows these two setups for a reducer of focal length f enlarging 0.5 × .

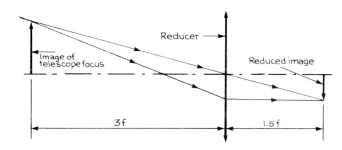

Figure 3-11.

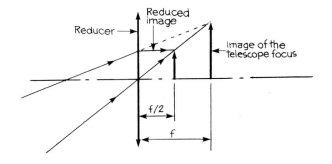

III.8. — THE CHOICE OF SYSTEM

We have just described five possible systems for taking astrophotographs: prime-focus, negative projection (Barlow lens), afocal, positive projection, and the use of a reducer. Each of these setups suits a specific situation. We are now going to explain in more detail their particular uses.

Of course, it is the focal length of your telescope that will determine what magnification is needed to achieve your effective focal length. The approach is always the following. I have a telescope of focal length F_o (for example, 900mm). I want to take a picture with an effective focal length F_e (for example, 4500mm). I must therefore achieve a magnification $\mu = F_e/F_o$ (here 4500/900 = 5). What optics must I choose to obtain μ?

Clearly, prime-focus photography is best when we desire an effective focal length close to the focal length of the telescope. Likewise, the choice of a shorter focal length suggests the use of a reducer. On the other hand, the Barlow lens is the simplest of the systems, and thus the most recommended, for magnifications of 2 or 3.

In fact, it seems from the first analysis that each level of magnification easily corresponds to one of the systems cited. Remaining are the comparable afocal and positive projection systems for magnifications greater than 3.

In every case you must use the simplest optical system. In fact, each optical element in the system absorbs light and introduces its own aberrations; therefore, you must limit the number of optical elements. This criterion leads us to choose positive projection. It is good to remember, however, that an optical system (an eyepiece, an objective) can be well-corrected for aberrations only for a given use, that is, for a certain position of the object.

Photographers are familiar with the following example. We can, with the help of an extension tube or a bellows, take photographs of very close objects (a few centimeters) with a standard 50mm lens. In addition, there are lenses of the same focal length, called "macro" lenses, specially designed for close-up photography, but equally useful for landscapes or, more generally, for objects at infinity. But why design two different lenses whose uses seem identical? The reason is quite simple. The standard lens is well-corrected for a distant object and will perform best in that situation; on the other hand, for close-up photography it will give poorer images than the macro lens.

The eyepiece is designed to give an image at infinity, that is, for observing an object in its focal plane, which is its normal visual use. The further removed the conditions of use are from the normal, the more the final image risks being altered. We should therefore use positive projection when the focus of the eyepiece is relatively close to the primary image, that is, to the focus of the telescope. We see in the Figure 3-12 that this case corresponds to a higher magnification.

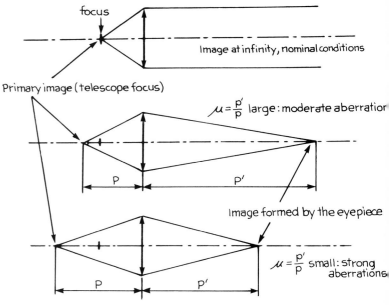

Figure 3-12.

One of the easily detected aberrations that appear when the eyepiece is used under bad conditions is the failure of the image to fall within one plane. On a picture of the moon taken with a magnification of 2 to 3 using positive projection, we see quite readily that it is impossible to obtain a sharp focus for the center and the edges of the field simultaneously.

For magnifications greater than about 6, it is generally agreed that the aberrations caused by positive projection become less than those created by the addition to the eyepiece of a camera lens. Thus, positive projection is the preferred system for this level. For smaller magnifications, on the other hand, it is recommended that you use an afocal system if it is of good quality. A system using the camera lens is preferable when the body is mounted with the eyepiece on an independent support and in line with the telescope's eyepiece. This is used when the telescope mounting is too weak to support the weight of the camera or to kill the vibrations caused by the shutter.

All these conditions lead us to Table 3-1, which supplies for each desired magnification the recommended system, the formula for the effective focal length F_e (for a focal length F_o of the telescope) and their most frequent uses.

Notice that with the systems using an eyepiece it is possible to add a Barlow lens in front of it, as is done in visual observations, to double the magnification. For enlarging the image, using a $2 \times$ Barlow lens in front of an eyepiece of a certain focal length is equivalent to employing an eyepiece of half the focal length, everything else being equal. For example, we can replace a 10mm eyepiece with a 20mm eyepiece mounted behind a $2 \times$ Barlow—the effective focal length will be the same.

In most cases we are not interested in making this substitution, since the faults of the Barlow lens must alter the quality of the final image. This is particularly true for the poor-quality Barlow lenses that often come with bottom-of-the-barrel telescopes.

There is an additional problem which is rarely considered. Suppose that we want a magnification of 8 or an effective focal length of $8F_o$ with a 10mm eyepiece used without a camera lens. We now replace this 10mm eyepiece with a 20mm eyepiece preceded by a Barlow. We always obtain an effective focal length of $8F_o$ and apparently nothing has basically changed. But the 20mm eyepiece itself perceives the combination ahead of it—objective of focal length F_o plus Barlow lens—as an objective of focal length $2F_o$. Thus, it behaves like an amplifying system of magnification 4 ($8F_o = 4 \times 2F_o$) and so we know it is working under bad conditions!

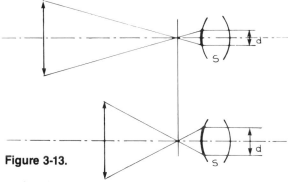

Figure 3-13.

In only one instance can the use of a Barlow lens with an eyepiece be beneficial, whether for visual observation or photography. The Figure 3-13 shows that the diameter d of the beam of light penetrating the lenses of the eyepiece (here in the case of a point object at infinity along the axis) is inversely proportional to the ratio F/D (focal length divided by the diameter) of the objective. Now, the greater the surface area of the eyepiece lenses, the greater the chances that the image will be deformed. It is for this same reason (among others) that photographers seek to stop down their lenses. As far as we are concerned, stopping down the entrance of the eyepiece is out of the question, since it would wind up losing a precious quantity of light. It is easy to see from Figure 3-14, however, that using a Barlow lens permits us to halve the diameter of the beam of incident light and thus—in the case of a very wide primary beam (Newtonian telescope with F/D = 4 to 6)—lets it strike the eyepiece better. This becomes more useful the more mediocre the quality of the eyepiece (Huygens, Ramsden, or Kellner, for example).

TABLE 3-1

μ	System	Principle	Effective focal length	Faint objects	Moon, Sun in their entirety	Details, Moon, Sun, Planets
				Use		
0.5	Reducer	(diagram) film, p'	$F_e = F_0 \left(\dfrac{p'}{f} - 1\right)$ (in general: $F_e = \dfrac{F_0}{2}$)	▮		
1	Prime Focus	(diagram)	$F_e = F_0$	▮	▮	
2 to 3	Barlow	(diagram) p'	$F_e = F_0 \left(\dfrac{p'}{f} + 1\right)$ (in general: $F_e = 2F_0$ or $F_e = 3F_0$)		▮	▮
3 to 6	Eyepiece & Camera Lens	(diagram)	$F_e = F_0 \times \dfrac{F_c}{l}$			▮
6	Eyepiece alone	(diagram)	$F_e = F_0 \left(\dfrac{p'}{f} - 1\right)$			▮

F_0 = Prime focus of telescope ; f = focal length of eyepiece, Barlow or reducer ; F_e = focal length of camera lens

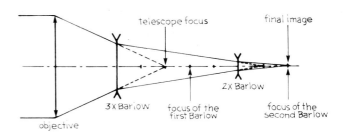

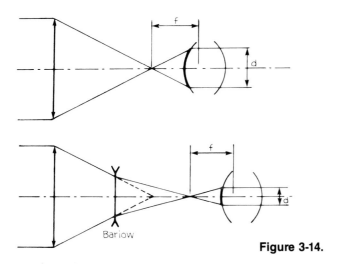

Figure 3-14.

Figure 3-15.

seems similar to the afocal system, since it provides magnifications of 4 or 6. For this level of magnification, professional astronomers almost always use negative projection, whereas the amateurs use practically nothing but the afocal system, which is the reason why we have covered it. Nevertheless, we judge that a Barlow coupling gives better results, if only because of the fewer optical elements the light must pass through. But the Barlow lenses must be of excellent quality. Their price is therefore higher, particularly for the first, which must have a larger diameter than the other. It is easy to understand why amateurs hesitate to spend the money, when they often have the necessary camera lens and eyepiece.

If the Barlow lens is itself practically exempt from aberrations, if the ratio F_o/D is small (less than 5 or 6), and if the desire magnification is higher (greater than 12), it can be beneficial to replace the eyepiece in the enlarging system with a combination Barlow lens and eyepiece. By chance, it is precisely with telescopes of small F_o/D that we have to use strong magnification. For instance, suppose that we wish to photograph a planet with an effective focal length F_e such that $F_e/D = 90$ (where D is always the diameter of the telescope). If F_o/D is 15 (in the case of a refractor) the necessary magnification will be 90/15 = 6. If F_o/D is 5 (in the case of a Newtonian reflector), the magnification becomes 90/5 = 18.

We have seen the case when a Barlow lens is added in front of an eyepiece, but why not add one in front of another Barlow lens? We thus create a negative amplifier made up of two successive Barlow lenses.

Figure 3-15 illustrates the use of a first Barlow lens of magnification 3 × followed by a second Barlow of magnification 2 × . In this case the total magnification is 6. This new system, using two Barlow lenses, thus

CAPTIONS FOR THE COLOR PHOTOGRAPHS

1. This photograph shows the rotation of the sky about the North Celestial Pole. Laurent Fournet took this exposure with a 50mm f/1.7 lens for 27 minutes.

2. With his camera fixed to a tripod Laurent Fournet took this exposure of the constellation Orion through a 50mm f/1.7 lens using Ektachrome 400 film exposed for 5 minutes. The star trails correspond to the sky's apparent displacement. Note the color differences among the stars.

3. Laurent Fournet took this photograph of the Milky Way in Scutum and Sagittarius using a 50mm f/1.8 lens. Clearly shown is M16, M17, M8, M20 and the trail of a passing satellite!

4. The Milky Way in the constellation Cygnus taken with a 50mm f/1.8 lens on Fujichrome 400 for 20 minutes. Note that the nebulae NGC 7000 faintly recorded. Herve Le Tallec mounted his camera "piggyback" on a 200mm telescope which he used for guiding.

5. M42 taken with the 60cm telescope at Pic du Midi by members of the Astronomy Club of the I.N.S.A. of Toulouse. The exposure was for 10 minutes on Ektachrome 400 with a F/D = 3.5.

6. Here the characteristic "redness" of some lunar eclipses has been recorded in the eclipse of 9-16-78 by Christian Viladrich. He used a 200mm Schmidt-Cassegrain telescope equipped with a focal reducer (F/D = 5) and Ektachrome film exposed for 10 seconds which was then processed for 640 ASA.

7. This photograph of the Moon in the third day of lunation by Jacques Silvain shows an over exposed crescent lit by the Sun with a properly exposed disk lit by light reflected from the Earth (Earthshine). It was taken with a 308mm telescope, F/D = 5 on Kodachrome 64 for 20 seconds.

8. Jean Mosser took this picture of M8 and M20 nebulae in Sagittarius with a Schmidt camera 210/310/495 (corrector plate diameter, mirror diameter, and focal length respectively, all in millimeters).

9. The galaxy M31 in Andromeda. Note that the periphery of the galaxy, where the young stars are, is more bluish than the center. Taken with a Schmidt camera on Fujichrome RD 100 film and exposed for 25 minutes.

10. Jean-Marie Roques took this picture of nebulae M16 and M17 with a Schmidt camera 210/310/495 on Fujichrome RD 100 film and exposed for 22 minutes.

11. The area of Orion's belt showing several red diffuse nebulae, with M42 at the right side of the field and the Horsehead Nebulae lower center left. Taken by Jean-Marie Roques with a Schmidt camera 210/310/495 on Fujichrome RD 100 exposed for 30 minutes.

12. Dr. Hans Vehrenberg took this tricolor picture of the Large Magellanic Cloud in the Southern Hemisphere using a 355mm diameter Schmidt camera. The tricolor print was made by combining three negatives of 103aO (7 minute exposure in blue light) 103aG (10 minute exposure in green light) and 103aE (15 minute exposure in red light).

13. The nebula NGC 7000 (North American) in Cygnus taken by Dr. Hans Vehrenberg using the tricolor technique with a 355mm diameter Schmidt camera. The tricolor print was made by combining three negatives of 103aO (7 minute exposure in blue light) 103aG (10 minute exposure in green iight) and 103aE (15 minute exposure in red light).

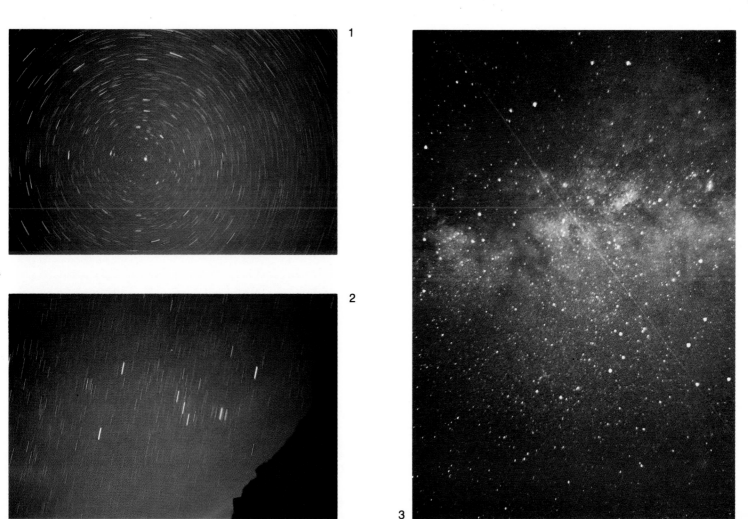

1

2

3

4 5

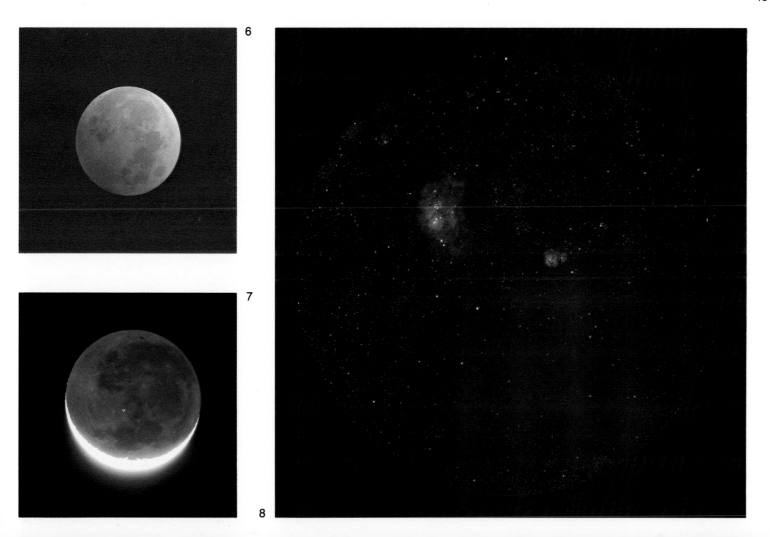

6

7

8

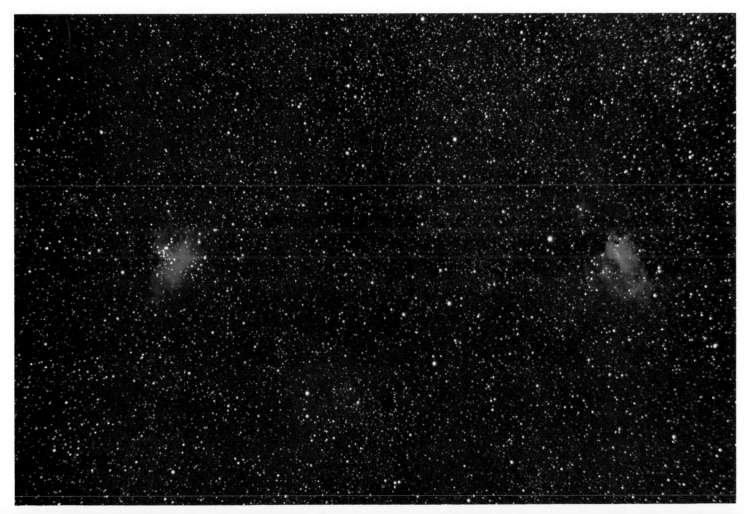

CHAPTER IV
FILMS, DEVELOPERS, AND FILTERS

IV.1. — CHARACTERISTICS OF FILMS

A photographic film is a surface sensitive to light. It consists primarily of a base (material and format vary) on which is applied a thin layer of emulsion, consisting of silver crystals in gelatin. These crystals are called photosensitive, since the light modifies their internal structure. With a suitable chemical treatment, or development, it is possible to distinguish the crystals that have received light from those that have not been exposed: after development, the crystal—consisting of silver bromide, silver chloride, or silver iodide—is transformed into dark metallic silver if it has been illuminated. In the zones where the film receives a lot of light, all the crystals are blackened; while on the unilluminated zones very few of the crystals are blackened.

IV.1.1. — CHARACTERISTIC CURVES

Every black-and-white film can thus be characterized by its curve of response (or its "transfer function") that is, the density of the blackening with respect to the quantity of light received. This curve is called the "characteristic curve of the film."

First of all, we must define the dimensions used to trace this curve. The quantity of light I received on the film per unit of surface for a unit of time is called the "illumination." The "exposure" E is the quantity of light received by a unit of surface; thus, it is the illumination multiplied by the duration of the exposure t: $E = I \times t$.

When a film is exposed, we can, after development, characterize its blackening by the proportion of light that it lets pass. We designate the *transmission factor* T of the film as the ratio of the intensity of the light passing through the film to the intensity of the incident light (Figure 4-1).

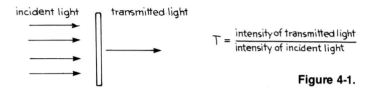

$$T = \frac{\text{intensity of transmitted light}}{\text{intensity of incident light}}$$

Figure 4-1.

We call the inverse of this transmission factor (that is, the value 1/T) the "opacity" of the film, but the value most often used is the decimal logarithm of the opacity, which we designate the "density of the film":

$D = \log 1/T$

Density depends upon the manner in which it is measured. Film tends generally to diffuse part of the incident light, and the measure of the density varies according to whether you consider the quantity of light transmitted along the axis of the incident ray (specular density) or integrate all the light transmitted by the film ("direct" light + diffused light). The proportion between the specular density and the diffuse density, called the Callier coefficient, is usually of the order of 0.7. Eye estimations of density are of the specular type, whereas the measurements obtained by laboratory densitometers are the diffuse type.

The characteristic curve must thus express a magnitude describing the state of blackening of the film as a function of the light received, that is, of the exposure. Since it is practical to use logarithmic units, the common practice is to represent the density as a function of the log of the exposure (Figure 4-2).

We see right away that the characteristic curve of a film can be broken down into five zones. Point 1 represents the film's threshold of sensitivity: at lower exposures, no image is formed and we say that the film is underexposed. On the other hand, there always exists a certain number of grains that, though not exposed, have been developed, which constitutes a background fog. Since in addition the base is not entirely transparent, we speak of the *fog level* of the film. The zones between points 1 and 2 and between points 3 and 4 are called the *toe* and the *shoulder* of the curve, respectively. Only in the central portion of the characteristic curve—between points 2 and 3—does a linear relation exist between the quantity of light received by the film and its blackening. Beyond point 4 the film is completely blackened, whatever the light received: we say that it is totally saturated or overexposed.

An image can form only when two different exposures cause different blackenings, that is, when the exposures lie between points 1 and 4 on the characteristic curve. A short time ago, we would consider a photograph "well-taken" when the exposure fell within the straight-line portion of the curve. Today, considerations of "signal-to-noise ratio," which we shall explain further on, tend to favor the top part of the toe of the curve.

IV.1.2. — FILM SPEED

We have just seen on the characteristic curve that the quantity of light that a film must receive to be correctly exposed is relatively well defined. The *sensitivity,* or *speed,* of a film is the concept which permits us to know the exposure it requires.

A large number of units of measurement for sensitivity exist. Their definitions are generally quite complex. Fortunately, they are relatively equivalent and are never used in an absolute fashion. Indeed, apart from the definition of the unit of sensitivity, we never calculate the exposure necessary for a film of given speed, but always proceed by comparison. For example, knowing that a film of a particular speed is correctly exposed with a particular known exposure, we can deduce that a film twice as sensitive requires half the exposure.

The most commonly used scale for characterizing the speed of ordinary films is the ASA scale, which enjoys the advantage of being linear: a film of ASA 200 is twice as sensitive as a film of ASA 100, and so on.

On the characteristic curves in Figure 4-3, film A is more sensitive than film B because it requires less exposure, or, with equal exposure, gives a denser image.

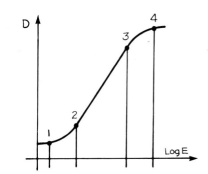

Figure 4-2.

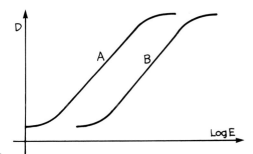

Figure 4-3.

Through a slight error in usage, sensitive films have come to be called "fast," since for the same subject they permit short exposure times; likewise, less sensitive films are known as "slow films."

IV.1.3. — CHROMATIC SENSITIVITY

The sensitivity of photographic films is not constant along the entire visible spectrum. The emulsions tend generally to be more sensitive to the short wavelengths than to the long. Whereas the eye does not see ultraviolet rays, they affect films quite easily.

Most current films, however, are corrected to have a rather flat response curve with regard to wavelength. These films are called "panchromatic." You can also find emulsion—called "orthochromatic"—which are insensitive to the long (i.e., red) wavelengths of the visible spectrum (Figure 4-4).

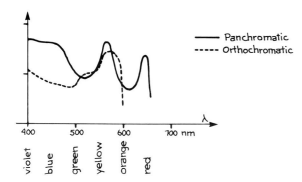

Figure 4-4. The chromatic sensitivity of Agfapan 25 (panchromatic).

There exist as well "infra-red" films, whose spectral sensitivity encompasses this radiation to about 1 micron. Used with a dark red filter (such as Wratten 29, which absorbs all wavelengths less than 6000 A), they allow you to take photographs in infra-red light.

Finally, we mention the Kodak scientific films (notably the 103a series) with restricted spectral sensitivity, which we shall come back to later.

IV.1.4. — THE CONTRAST OF A FILM

A film is called *contrasty* when it provides blackenings of very different intensities for a given interval of exposure.

The contrast of a film is characterized by the slope of the straight-line portion of its characteristic curve. This slope is called the γ (gamma) of the film (Figure 4-5). On the characteristic curves in Figure 4-6, film A is higher in contrast than film B ($\gamma A > \gamma B$).

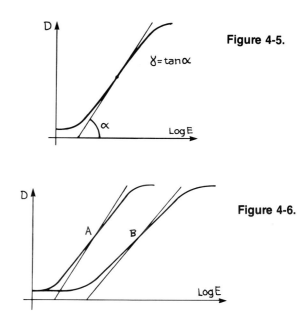

Figure 4-5.

Figure 4-6.

The higher in contrast a film is, the more apt it is to bring out details of neighboring luminosity, which is common in planetary photography. On the other hand, a contrasty film also presents an inconvenience: the range of exposure for which it is correctly exposed is narrower—you can have certain portions of the object photographed overexposed while other portions of the same object will be underexposed in the same picture.

IV.1.5. — RESOLVING POWER

One of the most important characteristics to know about a film when you want to do high-resolution photography is its ability to register small details. Indeed, each film is incapable of resolving images whose size lies below certain limits.

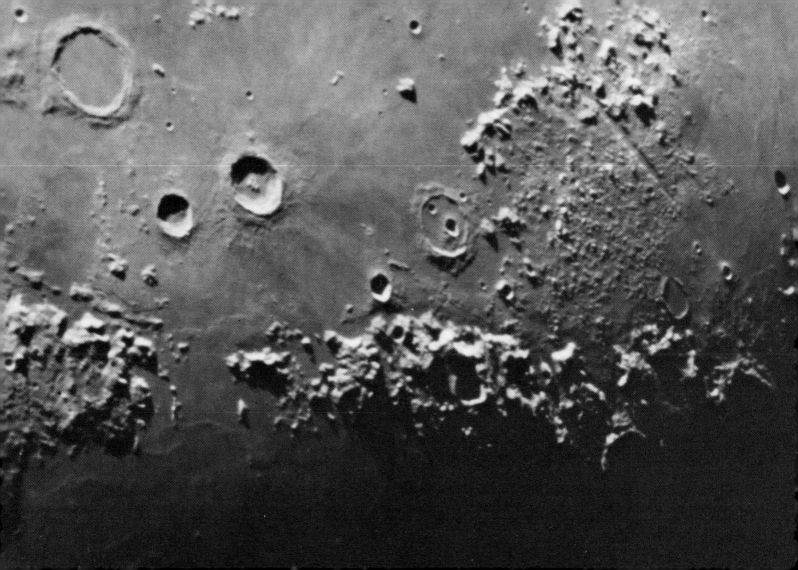

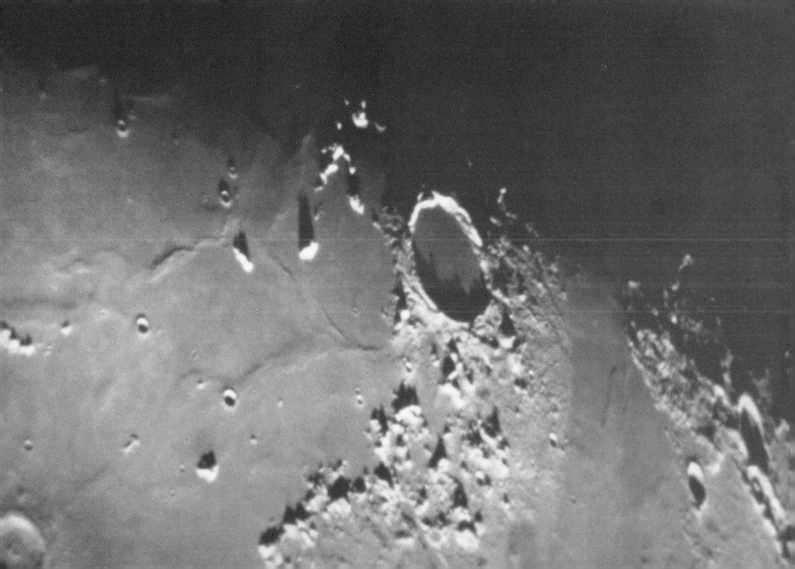

Photograph 4-1. This last quarter photograph (preceding left hand page) of the Alps region of the moon was taken by Jean Dragesco using a 355mm telescope, F/D 400 Kodak 2415 film with an exposure of 2 seconds. Compare this photograph with the one on the facing page.

Photograph 4-2. Alps region (preceding right hand page) on the moon during the first quarter taken by Jean Dragesco using a 355mm, F/D = 40, on Ilford XP1 film exposed for 1 second. Note that XP1 film has less contrast and resolving power when compared with Kodak 2415 (opposite).

The test typically used to evaluate a film's capacity to register small details consists of photographing targets made up of alternating black and white lines. Since the spacing of these lines varies, the film gives an image of a succession of lines closer and closer together. Beyond a certain limit—attained, for example, when the image contains 100 lines per millimeter—the film is incapable of distinguishing them because they are too close together. Thus, we say that the film resolves 100 lines per millimeter; it is capable of seeing a detail of 10 microns—this is what we call its *resolving power.*

The least powerful films do not resolve beyond 30 to 40 lines per millimeter (or 25 to 30 microns). By contrast, certain films specially designed for high resolution are made to resolve 400 or 500 lines per millimeter.

The concept of resolving power is more useful in a relative fashion—for comparing different films to one another—than as an absolute value for knowing a precise limit of resolution for each film. In fact, resolving power depends upon the form that the detail takes (series of dashes, single line, circular blots, etc.) upon the contrast, and . . . upon the accepted criterion for considering whether a particular detail is sufficiently visible or not in order to be considered resolved. Thus specialists in aerial photography know that it is common to see in high-altitude shots the white center line of a road when the width of this line is well below the theoretical resolving power of the film. Under the same conditions a series of parallel lines spaced at the same width would have given only a uniform gray spot (blurring of the image). A dark line on a white background or a line painted light gray (inverse contrast or weaker contrast) or a circular spot even two or three times wider in diameter than the line (different form) would not have

Photograph 4-3. Le Fur photographed the same lunar region as Photographs 4-1 and 4-2 using a 200mm telescope. It was also taken with the same 14 meter focal length which lets us note that the film used — Ilford FP 4 developed in Microphan — has less resolving power than XP1 or 2415.

been visible. Reciprocally, in planetary photography one is often disappointed to discover that details visible to the eye and whose size on the film is greater than its resolving power are indistinguishable on the photograph—in general, this is because they have too weak a contrast.

Another example will show to what extent you must be judicious with the concept of resolving power. Consider two films: the first has a rather large grain but a large gamma; conversely, the second presents a finer grain with a low contrast (note that such a difference can be obtained with the same film, developed in a high-contrast developer for the first case and in a low-contrast developer for the second). If the comparison test takes place with a contrasty target, the second will perform better, since the films are limited by their grain. On the other hand, with an object low in contrast, the first film will distinguish details that on the second will be drowned in the gray.

Moreover, the grain of the film greatly deforms details on the photographic image which are at the limit of the film's resolution. Thus, in order to obtain a good discernibility of image, you must anticipate a coefficient of at least 2 or 3 between the resolving power of the film and the finest details that you wish to register.

In sum, the notion of resolving power can be used either in a precise way for comparing films tested under rigorously identical conditions close to the conditions of use for which they are intended, or as an order of magnitude for determining the size that the smallest details you hope to register must have in the image.

To conclude this section, we shall prove the need for an enlarging system in high-resolution photography. Let F and D be the focal length and diameter of a telescope. Its resolving power is α ($''$) = 120/D for D in millimeters. The size at the telescope focus of an object of angle α is a = F $\times$ α (rd); after converting seconds of arc into radians, we find that the size at the focus of an object corresponding to the resolving power of the telescope is, in millimeters: a_o = 1/1720 $\times$ F/D. The image of smallest detail visible in a Newtonian telescope open to F/D = 6 is thus about 3 microns at the focus. No film useable in astronomy is capable of registering images this fine.

Photograph 4-4. *The craters Catharina, Cyrillus and Teophilus by Christian Arsidi at 203mm with a resultant focal length of 22 meters using Ilford HP 5 film. Like Tri X, HP 5 is a very sensitive but it also has fairly large granularity which appears in this photograph.*

IV.1.6. — A MATHEMATICAL MODEL

Aside from these notions, derived from common photographic practice, there have recently appeared concepts which are more mathematical and more specific to astronomical photography, derived from signal theories. They are relatively little known and few data exist about films available to the amateur. Nevertheless, they reach particularly interesting conclusions concerning the choice and utilization of films for astrophotography. This is why we have chosen to introduce them into this book. The reader who is little attracted to this somewhat abstract presentation, however, can content himself with the more concrete ideas previous, which we have just developed for his sake.

IV.1.6.1. — GRANULARITY

We know that photographic film consists of silver crystals which, once developed, form so many grains of metallic silver. When you photograph a uniform object, the image you obtain will also seem uniform—but greatly enlarged, it no longer seems so: "you can see the grain." The larger the grain of a film, the less it correctly restores the image.

For the subjective notion of graininess, professional photographers substitute the concept of *granularity,* which is a measure of the variations in density (notated σ D) within a uniform area. This concept directly interests the professional astronomer, who when he measures the dimensions of a faint nebula, for instance, wants to do so with the maximum of precision. If the grain makes the measured density constantly fluctuate, the precision will be poor; however, if the grain is very fine, the film gives an impression of a continuous relief and the measure obtained will be precise. In addition, the film's capacity for detection also depends upon it: if the grain is large, you can not detect a faint star; if the grain is fine, that star is visible. Note that this detection capability also depends upon the film's contrast: the higher the contrast, the better you can distinguish a faint object from a continuous background.

IV.1.6.2. — THE SIGNAL-TO-NOISE RATIO

The detection capability of a film is its ability to detect a faint "signal" buried in the photographic "noise," this noise being measured by the film's granularity. The signal-to-noise ratio is the objective measure of this detection capability. After what we have seen previously, it is not surprising to discover that, apart from a constant, it is expressed as the ratio of the contrast σ of the film to the granularity γ D of the film for the density for which it has been exposed:

$$R_{S/N} = \gamma / \sigma \ D.$$

For a given film the contrast varies with the density: on the characteristic curve, it is low at the beginning, increases along the toe of the curve, remains constant the length of the straight-line portion, then decreases after that (recall that the contrast is simply the slope of the characteristic curve). The granularity of the film is a function increasing with its density (we can show that it varies as the square root of the density). Consequently, the value of the signal-to-noise ratio will increase as the exposure of the film increases along the bottom of the characteristic curve, since at the beginning of the toe the contrast increases more quickly than the density and thus more quickly than the granularity. The ratio will then decrease along the straight-line portion, since the contrast there is constant while the density continues to increase.

When we superimpose the signal-to-noise ratio curve upon the characteristic curve of the film (Figure 4-7), we see that the former presents a marked bell-shaped curve whose maximum is generally located at about a diffuse density of 1 to 1.2 above fog level. It is at this density that the film's capacity for detection is greatest. Thus, it is at this density that you have to expose stellar photographs for which you wish to obtain the maximum information, as in the detection of very faint nebulae or high-magnitude stars.

Figure 4-7.

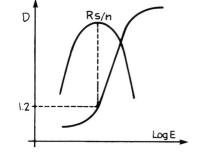

IV.1.6.3. — EQUIVALENT QUANTUM EFFICIENCY

This value characterizes the film's efficiency: it is the film's aptitude for giving a good signal-to-noise ratio in a minimum amount of time. Apart from a multiplicative constant, it is defined by:

$$EQE = (\gamma / \sigma \ D)^2 \times 1/E,$$

where γ, σ D and E are the contrast, the granularity, and the necessary exposure, respectively. The term 1/E characterizes the sensitivity of the film at the density where the EQE is measured, the sensitivity being the inverse of the quantity of light necessary to obtain this density while $(\gamma / \sigma \ D)^2$ is the square of the signal-to-noise ratio of the film at this density.

When we superimpose the equivalent quantum efficiency curve of the film upon its characteristic curve (Figure 4-8), it appears as a bell curve whose maximum is generally located around the diffuse density 0.6 above fog level. Thus, it is at this density that we must expose photographs when we desire the film's highest efficiency, particularly for high-resolution photos.

Figure 4-8.

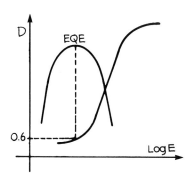

Notice that, because of the term 1/E, the maximum of the EQE curve lies in a region of exposures (thus of densities) lower than the maximum of the signal-to-noise ratio curve. This means that the same film will have to be less exposed when you desire high efficiency (high resolution) than when you favor the signal-to-noise ratio (detection of faint objects).

Although the concepts of signal-to-noise ratio and equivalent quantum efficiency are quickly demonstrated mathematically, we prefer here to remain at the concrete level and stress their practical importance. The signal-to-noise ratio characterizes the film's capacity for detection, such as its success in registering a faint star above a nebulous background or low-contrast details on a planetary surface—but also in obtaining with the maximum sharpness a series of black-and-white lines (photographic target) whose spacing is comparable to the size of the film grain. In fact, the resolving power is strongly related to the signal-to-noise ratio.

So you must not be surprised that the square of the signal-to-noise ratio appears in the expression for equivalent quantum efficiency. Consider planetary photography, for example. For reasons we shall supply later, we must try to expose the film as rapidly as possible. We use the telescope with an amplifying system (focal length) such that the scale at the focus (that is, the size of the image in seconds of arc per millimeter) makes the resolving power of the instrument compatible with the resolving power of the emulsion used (in fraction of mm). The classic dilemma confronts us: in order to have the shortest exposure time, is it better to use a fast film with a long focal length (which is necessary because of the poor resolving power of fast films) or a fine-grain, but slow, film with which we can use a shorter focal length? The size of the image (thus the film's resolution) influences the square of the exposure time (because by using a focal length twice as short we concentrate the same quantity of light into four times less surface) while the sensitivity is inversely proportional to the exposure time (if the latter is short). To optimize the exposure time, we must thus use the film for which the product (resolution)² × sensitivity is the highest. Now, this expression is no more than a disguised way of writing the equivalent quantum efficiency.

We have seen the problems posed by the objective determination of a film's resolution. It depends, in fact, upon the form, the contrast, and the density of the target used, and . . . upon the person doing the measuring. The equivalent quantum efficiency, though, is an objective measure of the film's quality.

IV.2. — THE DEVELOPER

IV.2.1. — GENERALITIES

A film's characteristics are not immutable; they are influenced by the conditions of development. In particular, the choice of developer and the duration of development can modify the film's speed, contrast, and grain.

We can classify as "normal" those developers which give a film its nominal speed. These are the developers most often used in standard photography—Kodak's D76 and Tetenal's Ultrafin, for example.

The aim of a low-contrast developer (such as Kodak's Microdol) is to diminish the film's granularity and thus improve its resolving power. In return, the speed and the contrast are reduced. Perceptol, manufactured by Ilford, reduces the nominal sensitivity of the film by as much as one half.

On the other side, energetic developers permit an increase in the film's speed as well as in its contrast. Unfortunately, this is always accompanied by an increase in grain. Microphen (Ilford) and DK50 (Kodak) bring a gain in speed of about 50%. D19b (Kodak) and Acuspeed (Paterson) permit you to triple the speed of certain films (Kodak's Recording Film, from a nominal speed of ASA 1000, can even be pushed up to ASA 6400 with a normal development in Acuspeed).

An increase in development time accomplishes the same thing as using a more active developer: the speed and the contrast of the straight-line portion increase, but so does the grain. This increase in grain is brought about by a greater density of the fog and is equally responsible for a certain decrease of contrast along the toe of the curve.

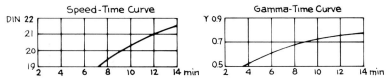

Figure 4-9. The influence of development time upon the characteristics of the film (in this case, developer: Rodinal; film: Agfapan 100).

IV.2.2. — RECOMMENDED DEVELOPERS FOR ASTRONOMY

IV.2.2.1. — MWP-2

The name of this developer, composed of the initials of one of the most prestigious observatories, "Mount Wilson and Palomar," indicates its origins. For the amateur astronomer it has the inconvenience of not being commercially available; thus, you must make it yourself.

To make this developer, dissolve in the order given the following ingredients in 750ml of distilled water at 50°C:

sodium sulfite	105g
hydroquinone	10g
phenidone	0.4g
benzotriazole	0.6g
potassium bromide	2g
potassium carbonate	30g

Complete by adding cold water to make a liter. Developing time is between 7 and 12 minutes. Additional information about this developer can be found in *Astronomical Journal*, vol. 73, no. 9, Nov. 1968.

MWP-2 gives optimal results for spectroscopic films like the 103a series, for which it is appreciably superior to D19b.

IV.2.2.2. — D19b AND ACUSPEED

These two developers, similar in performance, are among the most energetic that exist: they furnish higher speeds and a deep contrast. They are the ones most frequently used in amateur astronomy. In comparison to MWP-2 they have the advantage of being available at major retailers of photographic products. D19b is a powdered developer made by Kodak. Acuspeed, marketed by Paterson, comes in the form of a liquid concentrate.

IV.2.2.3. — HC110

This is a liquid developer from Kodak which, just like Acuspeed, has the advantage of letting you prepare it just before use (you must always let a powdered developer sit after mixing one to two days before use). HC110 is not a miracle developer, but it enjoys a great flexibility and accommodates itself very well to Kodak 2415 film. Indeed, there are six standard dilutions which let you modulate the film's contrast: dilution A (1 + 15), B (1 + 31), C (1 + 19), D (1 + 39), E (1 + 47), and F (1 + 79).

IV.2.2.4. — LOW-CONTRAST DEVELOPERS FOR 2415

As we shall see later, Kodak 2415 is an invaluable film for astronomy; however, it naturally has a very high contrast. Now, in certain cases such a contrast can be detrimental (a picture of the moon, for instance). If you wish to employ 2415 in those cases, you need to develop it in a special developer which lets you significantly diminish the contrast.

For some time astronomers have been familiar with Pota, a very low-contrast developer which gives 2415 a contrast comparable to that of ordinary films, or even lower. Unfortunately, Pota is not marketed and you will have to make it yourself according to the following formula:

distilled water	1 liter
phenidone (developing agent BD84 of Kodak or phenidone of Ilford)	1.5g
sodium sulfite	30g

Recently Kodak has put on the market another very low-contrast developer, Technidol LC, which gives 2415 results equivalent to those with Pota.

V.3. — THE PROBLEM OF LONG EXPOSURES

V.3.1. — RECIPROCITY FAILURE

Imagine that photographing a certain object requires an exposure time of 1/100th of a second. To photograph an object ten times less luminous will demand, under the same conditions, an exposure time ten times as long, or 1/10th of a second. This is called the law of reciprocity: the product $E = I \times t$ is constant.

Now, for the majority of ordinary films this law no longer holds as soon as the exposure time passes about 1 sec., and this *reciprocity failure* (also called the *Schwarzschild effect*) becomes more pronounced the longer the exposure. In this way, if an object requires a 1 sec. exposure, an object ten times less luminous can require 20 sec. (and not 10 sec.), an object one hundred times less luminous around 400 sec., and so on (Figure 4-10).

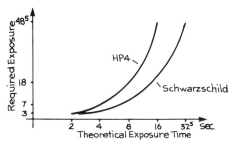

Figure 4-10. Schwarzschild's approximation states that the theoretical effective exposure for a long exposure of a film is the power 0.8 of the actual exposure. Here this law is compared to the values furnished by Ilford for HP4.

To understand the cause of this phenomenon, we must return to the elementary photochemical reactions taking place in the film's sensitive layer. The Ag^+ silver ions contained in this layer change state when they are struck by a photon (particle of light): they are transformed into a silver atom and in this way keep a record of the photon's passage. This new atom of silver, however, is relatively unstable in the gelatin and easily returns to its initial state (thus the information of the photon's passage is lost) unless then neighboring Ag^+ ions have themselves also reacted. Therefore, a sufficient number of reactions must take place, that is, a sufficient number of photons must strike the film during the average life of the isolated atom. In the case of faint-object photography, the flux of photons is not dense enough, which causes a diminution in the percentage of effective photons.

Note that ordinary films are more subject to reciprocity failure the higher their nominal speed (for low exposures); thus, the deviations in sensitivity among the different films tend to diminish when the exposure times increase. The speed of a fast film is reduced on average by a factor of ten at the end of a quarter hour of exposure. Moreover, reciprocity failure produces a decrease in contrast and in the maximal density of the film.

Reciprocity failure never bothers the majority of film users, since it intervenes only for exposures longer than a second, which is uncommon in conventional photography. By contrast, the astronomer is very sensitive to it, because in stellar photography the exposure time easily runs into minutes or even hours. Not only does the law of reciprocity require long exposures for faint objects (galaxies, nebulae, etc.) with all the problems that that entails, but the *failure* of the reciprocity law often requires that this exposure time be multiplied by 2, 3, or even more.

There are two solutions for countering this inconvenience: corrected films and hypersensitization.

IV.3.2. — FILMS CORRECTED FOR RECIPROCITY FAILURE

Kodak makes special films with practically no reciprocity failure for exposures less than an hour. These films are denoted in the catalogues by the letter "a": IIaO, 103aF, etc. The 103a films, for example, have the same speed as an ASA 80 film for short exposures, but this speed decreases little in spite of long exposures. Thus, a seven-minute exposure on 103a has the same result as 20 minutes on a traditional ASA 400 film.

IV.3.3. — HYPERSENSITIZATION

Hypersensitization consists of treating an initially uncorrected film before exposure so that its reciprocity failure is considerably diminished. This treatment is sometimes optical, but is more often chemical.

IV.3.3.1. — OPTICAL HYPERSENSITIZATION

You can easily get a rough understanding of the principle behind this by examining the characteristic curve of a film. In order for an image to be correctly visible on the negative, it must receive at least a quantity of light h_2 (Figure 4-11).

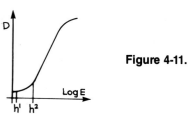

Figure 4-11.

Optical hypersensitization, known as *pre-exposure,* consists of uniformly exposing the film to an exposure h_1 a little before its astronomical use. This h_1 corresponds to the film's threshold of sensitivity, the fog that results being faint. On the other hand, it then takes only a quantity of light $(h_2 - h_1)$ to obtain an image of the object photographed, with a satisfactory contrast.

Pre-exposure interests amateurs because it does not require special materials. On the other hand, it is a delicate operation, since the quantity of light h_1 must be sufficient without running the risk of fogging the film. You must make several attempts to determine this quantity with precision.

It must be noted that pre-exposure always entails an increase in fog and a decrease in contrast. We thus do not recommend it for films already low in contrast or for the detection of objects whose contrast with their surroundings is low.

IV.3.3.2. — HYPERSENSITIZATION IN SILVER NITRATE

This operation takes place in three stages. First of all, you have to immerse the film in a solution of silver nitrate—the easiest way is to use a developing tank. The bath should last 6 minutes for a 0.5×10^{-3} molar solution. During these 6 minutes you should agitate the film regularly at the rate of a few turns per minute.

After the silver nitrate bath the film should be rinsed for 3 minutes in alcohol at 90°C then dried in a flow of dry air.

The film must be exposed and developed immediately after the silver nitrate treatment. The gain in speed obtained for exposures of about ten minutes or so is a factor of 4 to 5 for films such as Tri X, and even 8 to 10 for Kodak 2415. This gain becomes more appreciable the more sensitive the film is to red.

Less effective than treatment with "forming gas," the silver nitrate bath does, however, have the advantage of not requiring special equipment.

IV.3.3.3. — HYPERSENSITIZATION WITH FORMING GAS

Among the numerous chemical hypersensitization processes, treatment with forming gas is the most widely used by amateurs, since it is both simple and effective.

This process consists of treating the film in a mixture of nitrogen (92%) and hydrogen (8%), known as "forming gas." The film is placed in a lightproof tank, preferably wound onto a developing spool rather than left in its container. After producing a vacuum in the tank, the forming gas is introduced under pressure and heated to about $50°C$. The length of this treatment varies, according to the film, between 3 and 48 hours. After this time, you need only evacuate the gas and recover the film.

You should use the film within a few days after treating it, but the treatment is particularly effective.

The results of hypersensitization vary considerably from one film to another. Table 4-1 gives, for different films, the nominal speed (without hypersensitization and for short exposures); the length of treatment in forming gas; its period of effectiveness; the gain in speed that it produces, compared to the untreated film, for long exposures of 10 to 60 min.***; and finally, the relative speed, for long exposures, of all these films after treatment, the untreated 103aE being taken as a reference point.

Hypersensitization acts upon the film to increase its speed, but especially to reduce its reciprocity failure; thus, the gains from this treatment essentially concern long exposures. For brief exposures, the effect is less evident. Note, however, that 2415 already presents a gain of about 3x for exposures on the order of one to a few seconds.

IV.3.3.4. — OTHER HYPERSENSITIZATION TECHNIQUES

Forming gas is not the only possible means of hypersensitizing a film by heating it in a particular gaseous atmosphere. Air and nitrogen are also used. Moreover, storage in hydrogen can constitute a treatment of the film or serve as a preparation for another gaseous treatment. The same holds true for the evacuation technique itself which consists of completely removing gas from the film in the hypersensitization tank with a vacuum pump. This evacuation brings interesting gains by clearing the emulsion of the oxygen and water vapor it contains.

***In this experiment, Marling calls gain in speed the ratio of exposure times necessary for the same subject for an untreated film and for a hypersensitized film; it is more standard in sensitometry to compare the necessary illumination for equal exposure times.

TABLE 4-1
Gas Hypersensitization for Various Films

Film	Nominal ASA Speed	Hours of Treatment	Shelf Life in Days	Gain in Speed	Relative Speed
2415 (+ D 19 b)	100 ASA	36-48	– **	15 × to 50 ×	3.4
Tri X	400	15-18	5-7	5 × to 10 ×	1.4
HP 5	400	7-9	3-4	10 × to 15 ×	2.9
103aE (untreated)	*	–	–	–	1.0
103aE	*	4-5	1-2	2 × to 3 ×	2.9
103aF	*	5-6	1-5	2 × to 3 ×	4.0
103aO	*	6	2	2 × to 3 ×	3.7
Ektachrome 200	200	9-11	4	4 × to 10 ×	1.7
Ektachrome 400	400	6-7	2-5	4 × to 10 ×	1.6
Kodacolor 400	400	5-6	2	7 × to 12 ×	3.8
Fujichrome RD 100	100	5-6	2	4 × to 10 ×	1.7
Fujichrome 400	400	5	0.5 to 2	2 × to 8 ×	1.7

* One cannot determine a speed in the usual sense of the term for 103a films because of their particular chromatic response; a rough estimate for them might be around ASA 80.

** Not determined by Marling

Table Adapted From J.B. Marling, "Gas hypersensitization of 35mm films," Astrophoto, IV, Feb. 1981.

Besides the silver nitrate bath, it is possible to do "soakings" in distilled water or in an alkaline solution, particularly ammonia. This chemical treatment must be done very soon before exposure and is really effective only for films sensitive to red or infra-red (Kodak 2415 or Kodak films of special classes N and Z).

Finally, you can cool the film at the moment of exposure to about $-20°$ to $-30°C$. This can be done with the "cold cameras" which have appeared on the market . . . or without any effort during a winter observation at Pic du Midi.

Hypersensitization is a very recent technique, still little practiced by amateur astronomers. The next few years will permit us to accumulate the necessary experience in the choice of the most efficient methods and we shall no doubt see more specialized materials, easily accessible to amateurs appear on the market.

Photograph 4-5. The nebula NGC 7000 in Cygnus (preceding left hand page) by Dominique Albanese. Schmidt camera of focal length 228mm, F/D = 1.65 on 103aF film exposed for 25 minutes through a Wratten 25 red filter and developed in D19b.

Photograph 4-6. Here (preceding right hand page) Dominique Albanese used Kodak 2415 film hypersensitized in silver nitrate to photograph NGC 7000. All conditions of exposure and processing are identical to Photograph 4-5. Comparison of the two photos shows the effectiveness of treatment in silver nitrate for increasing red sensitivity.

Photograph 4-7. Compare this photograph by Bertrand Koehler with Photographs 4-5 and 6. Here Kodak 2415 was hypersensitized in forming gas and exposed using a 200mm focal length telephoto, F/D = 2.5 for 15 minutes. Note that the film here received 4 times less light than for the two preceding photos.

IV.4. — BLACK-AND-WHITE FILMS USED IN ASTRONOMY

The majority of amateurs use ordinary films, but special films are available and have application to astrophotography. Here we shall draw up an inventory.

IV.4.1. — COMMON HIGH-DEFINITION FILMS

These films, designed especially for reproduction, are characterized by a remarkable resolving power (a few microns) and a high contrast. Their disadvantage for the astronomer is their slow speed.

The best known of these films is Kodak's Recordak, whose speed unfortunately does not pass ASA 10.

In the same category you can also find Agfaortho 25, which (as its name indicates) is an orthochromatic film of ASA 25 marketed by Agfa. Of a lower resolving power but faster than Recordak, it has the additional advantage of being one of the only orthochromatic films that you can procure easily.

IV.4.2. — COMMON FILMS OF AVERAGE SPEED AND CONTRAST

We shall group under this heading all the films that photographers tend to use for landscapes, portraits, family or vacation photos, etc. They are characterized by a moderate contrast, permitting a certain latitude of exposure and a resolving power which, without being exceptional, is sufficient for conventional subjects. Their nominal speeds extend from ASA 25 to ASA 400.

As a whole, these films follow relatively well the law that the resolving power diminishes as the speed increases.

Table 4-2 lists the films of this category, classed according to their speed, for three of the principal manufacturers. All the films presented here are panchromatic.

TABLE 4-2
Commonly Available, Moderate Contrast Films

	Agfa	Ilford	Kodak
25-50 ASA	Agfapan 25 (25 ASA)	Pan F (50 ASA)	Panatomic (32 ASA)
100-125 ASA	Agfapan 100 (100 ASA)	FP 4 (125 ASA)	Plus X (125 ASA)
400 ASA	Agfapan 400	HP 4 HP 5	Tri X

IV.4.3. — COMMON HIGH-SPEED FILMS

The two best-known high-speed black-and-white films are made by Kodak: Recording Film (ASA 1000) in 35mm format and Royal X Pan (ASA 1250) in 120 format. These films are characterized by a high sensitivity, but unfortunately, by a high granularity and low contrast as well.

Although designed for dim subjects, they are rather sensitive to reciprocity failure. Because of this, they are not preferable to Tri X for long exposures in spite of a much higher nominal speed.

IV.4.4. — VARIABLE SPEED FILMS

In 1980 Agfa and Ilford simultaneously perfected two films of a new type. These are black-and-white films whose treatment, inspired by that of color films, permits professional laboratories to recover the silver contained in the gelatin. It must be said that soon before the perfection of these emulsions, the escalation in the price of silver had caused a corresponding leap in the price of film.

For the user the advantage of these films—Vario XL from Agfa and XP1 from Ilford—does not reside in the recovery of the silver, but in the fact that these emulsions tolerate a wide range of illumination. Their speed is said to be variable since at the same time one can use them as a slow film (by overexposure) or as a fast film (by underexposure).

Photograph 4-8. The craters Arzachel, Alphonsusand Ptolemaeus by Christian Arsidi using a 280mm telescope with an effective focal length of 16 meters on Ilford XP 1 exposed for 1 second.

Color films contain several sensitive layers, each reacting to a given color. The morphology of variable speed films is similar. They contain several layers, but instead of being selective, the layers have different sensitivities and are thus able among themselves to react over a very extended total range of illumination.

XP1 seems better adapted to astrophotography than its competitor, Vario XL. For a speed of ASA 400 it has a resolving power of 130 lines per millimeter, which is very good for this range of sensitivity.

IV.4.5. — KODAK 2415

This is a fairly recent Kodak film (late Seventies). Developed normally, it is characterized by a heightened contrast, a remarkable resolving power, and a speed sufficient for astronomical use. It is not surprising, then, that it has been greeted with enthusiasm in the small world of astrophotography.

Like all films, the characteristics of 2415 depend upon the development. In this case, the contrast can vary by very large proportions (γ = 0.4 to γ = 3.5). Table 4-3 indicates the gamma and the speed of 2415 according to the developer and the developing time. The asterisk indicates the most common developing time for each developer. By way of comparison, standard films developed normally have gammas ranging from 0.6 to 0.85.

The speed-resolution ratio is excellent for 2415. Table 4-4 compares the resolution in lines per millimeter of 2415 developed in HC-110 (dilution D, 8 min.) and in Pota (15 min.) to that of Panatomic (figures obtained from Kodak Rochester):

Although less sensitive, Panatomic clearly has a lower resolving power. We have seen that the equivalent quantum efficiency of a film can be expressed as the product (resolution)² × sensitivity. Applied to the Table 4-4, this criterion proves the incontestable superiority of 2415. Another advantage of 2415: is that it has a greatly extended chromatic sensitivity which permits its use from ultraviolet far into the red (Figure 4-12).

In conclusion, 2415 seems on the one hand like a film very well adapted to high-resolution astrophotography (thanks to its excellent level of speed-resolving power) and on the other hand like a very versatile tool, since its contrast is easily modulated according to its treatment and can thus be adapted to the optimal value required for each object. It is indeed the miracle film of our time which, used with an adapted developer, possibly hypersensitized, represents the best solution to ninety percent of an amateur astronomer's needs.

TABLE 4-3
Contrast and Speed for 2415 Using Varous Developers

Developer	Duration (min) at 20°C	Contrast (γ)	Speed (ASA)
Dektol	3	3.6	200
D19b	8	3.5	200
,,	4*	2.9	125
,,	2	2.8	100
HC 110 (Dilution B)	12	2.7	250
,,	8*	2.05	200
,,	6*	1.6	160
,,	4	1.4	100
HC 110 (Dilution D)	8*	2.0	125
,,	6*	1.65	100
,,	4	1.35	80
HC 110 (Dilution F)	10	1.3	64
,,	8*	1.2	50
,,	6	1.05	32
D 76	12	2.5	135
,,	10	2.0	120
,,	8*	1.5	100
,,	6	1.2	50
Pota	20	0.9	32
,,	15*	0.6	25
,,	10	0.4	25
Technidol LC	18	0.75	32
,,	15	0.6	25
,,	12	0.5	25

* Most common developing time

TABLE 4-4
Resolution (lines/mm) for 2415 Developed in HC-11OD and Pota

	2415 HC-110	2415 Pota	Panatomic
Speed (ASA)	125	25	32
High Contrast Target (1000 : 1)	320	400	200
Low Contrast Target (1,6 : 1)	125	125	80

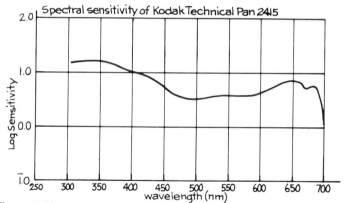

Figure 4-12.

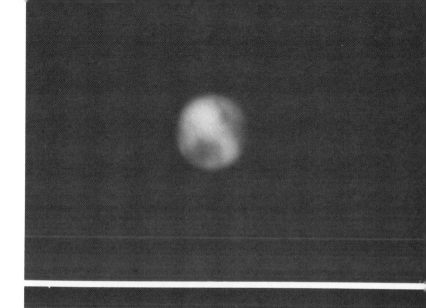

Photograph 4-9. *Mars (top right) photographed by Christian Arsidi using 280mm telescope with an effective focal length of 16 meters on Kodak 2415 exposed for 2 seconds.*

Photograph 4-10. *Jupiter (bottom right) photographed by Jean Dragesco using a 355mm telescope, F/D = 50 on Kodak 2415 film exposed for 2 seconds and developed in D19.*

IV.4.6. — KODAK 103a AND IIIaJ FILMS

For a number of years Kodak has made a range of films and plates known collectively as "spectroscopic." These are, in decreasing speed (thus in increasing resolution) the emulsions of type Ia, 103a, IIa and IIIa. The letter a, for "astronomical," means that these emulsions have a small reciprocity failure; thus they are appropriate for long exposures. Among them, the 103a and IIIaJ films are readily available to amateurs in 35mm format.

On a high-contrast target, the 103a films have a resolution of 80 lines per millimeter. Under the same conditions, IIIaJ resolves 200 lines per millimeter, but has only one sixth the speed.

The 103a and IIIaJ films should be developed in D19. This developer gives them a heightened contrast which increases for the 103a films according to the length of development:

for 2 min.: gamma = 1.1
for 4 min.: gamma = 2.0
for 8 min.: gamma = 2.7
for 12 min.: gamma = 2.9

For the same development times, the contrast for IIIaJ remains constant, but with a very high gamma (gamma = 4).

In general, you should choose a development of 4 minutes at 20°C, which represents the best signal-to-noise ratio for these films.

The Kodak spectroscopic emulsions come with different chromatic sensitivities. The spectral classes of these films are designated by a letter following the type denomination, in order to characterize the emulsion completely. For example: IIaO, 103aF, etc.

Photograph 4-11. The galaxy M101 (top right) photographed by Christian Arsidi using a 280mm Schmidt-Cassegrain telescope and focal reducer for a F/D = 5 on Kodak 103aO film exposed for 45 minutes.

Photograph 4-12. M101 (bottom right) photographed under the same conditions as Photograph 4-11 but with Kodak 2415 film hypersensitized in forming gas. Note that 2415 has better definition than 103a and that hypersensitization gives it slightly superior sensitivity for long exposures.

Figure 4-13 indicates the chromatic sensitivity of each class while the curves in Figure 4-14 represent the chromatic sensitivities of emulsions 103aE, 103aF, and 103aO, which can easily be procured in 35mm format.

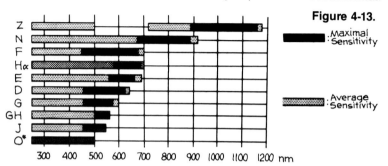

Figure 4-13.

⬛ : Maximal Sensitivity

▨ : Average Sensitivity

Figure 4-14.

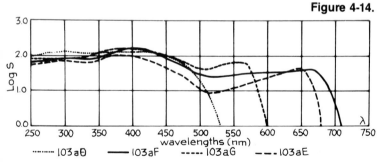

······ 103aθ —— 103aF ----- 103aG — — 103aE

IV.5. — COLOR FILMS

Until recently astronomers tended to use black-and-white films much more often than color films. There are many reasons why:

1) The performance of color films was more limited than that of black-and-white films; there still do not exist color films specially adapted to astronomy (high contrast fast films, films corrected for reciprocity failure).

2) The problem of processing. Color film development, laborious and delicate, is generally confined to a commercial laboratory. Thus astronomers lose the possibility of adapting the treatment to their needs and can not examine the results in the few hours following exposure, which is necessary to correct inevitable mistakes.

3) The price of color is higher than black-and-white. Astronomical photography is difficult and quite often you have to make numerous attempts before getting a successful picture.

4) The lack of fidelity. The colors rendered by a film are often inexact, especially for long exposures. Astronomers desiring a color photograph with scientific value often prefer to make exposures (usually three) on black-and-white film with interpose filters of different frequency ranges (usually blue, yellow and red); the colors are then restored (and adjusted) on printing paper in the darkroom. This technique is called the three color process and is discussed in Section VII.4.2.

5) Finally, color adds nothing to certain astronomical subjects. The moon and sun, for example, offer nothing more in color, even from an aesthetic point of view. On the other hand, color is imperative for lunar eclipses and can give spectacular results for nebulae.

However, since the beginning of the Eighties, much progress has been made in the chemistry of color emulsions (important improvements in speed and resolving power). The major manufacturers, lead by Kodak and Fuji, have in short time put on the market several high-performance films intended for the general public, yet of benefit to amateur astronomers.

Here we shall cover the films of greatest interest to astronomy.

Among the slow films, the place of honor has been occupied for many years by Kodachrome 25 and 64. Note that color films are designated by the name of the type of emulsion followed by a number indicating the speed in ASA. Terms in "-chrome" apply to reversal films, terms in "-color" to negative films.

Thanks to their remarkable definition, these films are prized for planetary photography. Their failing lies in the special processing they require, which means that they must be sent to a specialized laboratory. However, in 1984 Fuji revised all its reversal emulsions and put on the market Fujichrome 50D and 100D. Both have a resolution comparable to Kodachrome and a slightly higher speed, but enjoy the advantage of being compatible with the E6 processing available to amateurs.

We have seen that faint-object photography requires a fast film (if possible, without sacrificing too much resolution) that is not too susceptible to reciprocity failure or is highly responsive to hypersensitization. In choosing a color film, we must also make sure that its chromatic balance remains undisturbed for long exposures.

Until 1983 the speed for color films did not exceed ASA 400, and the best choice for a long-exposure photo was the reversal film Fujichrome 400. This film provides an excellent speed-resolution compromise, renders the background relatively neutral for long exposures and is fairly sensitive in the red where the emission line Hα for gaseous nebulae lies. In addition, it has been shown that you can increase the speed and contrast of Fujichrome 400 by developing it like a negative film in a C41 treatment. You can then duplicate this negative on Kodak 5072 reproduction film (using filtration adapted to the sources of light employed) in order to obtain a slide.

In 1983 two ASA 1000 emulsions appeared: the negative film Kodacolor VR 1000 and a reversal film manufactured by 3M. As can be expected, these emulsions show more granularity than ASA 400 films. Like its cousin Kodacolor 400, VR 1000 demonstrates a remarkable sensitivity in the blue, which makes it valuable for galaxies and certain nebulae. On the other hand, it is also very sensitive to red light, like Ektachrome 400, in a region of the spectrum where Kodacolor 400 exhibits a few weaknesses.

The 3M ASA 1000 film has two inconveniences: a high granularity and a greenish coloration of the background sky. To eliminate this coloration, you must reproduce the slides obtained by interposing a CC 30M (magenta) correction filter (of course, the magenta filter must not be used during the sky exposure in order not to increase exposure time). Hypersensitization in Forming Gas does not give good results with 3M 1000 (it creates a strong dominant green overcast.) On the other hand, you can push its speed to ASA 2000 without penalizing resolution too much; however, development at ASA 4000 (possible according to the manufacturer) causes strong granularity and bad color balance. Finally, 3M 1000 has a spectral sensitivity greatly extended toward the short wavelengths, attaining almost 3800 Å in the ultraviolet.

Another stage was reached in 1984 with the appearance of ASA 1600 films. First of all, Ektachrome 800/1600, which is in fact a professional film, can be developed at ASA 400, 800, 1600, or even 3200. Disappointing after the first attempts at nebulae photography, this film nevertheless has a remarkable speed-resolution ratio, especially when developed at ASA 1600. It should therefore find application in planetary photography with a long focal length.

Fuji offers two ASA 1600 films: Fujichrome RSP 1600 ProD in slides and Fujicolor HR 1600 in negatives. These emulsions seem to have a promising future in astrophotography, but at the time of this edition they are too new for us to form a definitive opinion.

IV.6. — FILTERS

IV.6.1. — WHAT IS A COLOR FILTER?

We know that a ray of light is characterized by its wavelength λ. This wavelength determines the nature of the light—indeed, this is not limited to the visible domain but comprises at one end infra-red rays, microwaves and radio waves, and at the other end ultraviolet rays, x-rays and gamma rays (Figure 4-15).

Visible light thus represents a very small band, extending from about 4000Å to 7000Å. In this band the eye is sensitive to light and perceives different wavelengths as different colors (Figure 4-16).

In fact, a beam of light is almost never monochromatic, that is, composed of rays all having the same wavelength. On the contrary, it generally contains rays of all wavelengths with, however, different intensities according to λ. We can thus describe its nature by presenting the curve of luminous intensity as a function of λ.

Thus, in Figure 4-17 the beam of light A will be more greenish and the beam of light B more reddish.

A filter is made from a substance which absorbs part of the light passing through it. We can characterize a filter by its coefficient of transmission T, which is the percentage of light transmitted by the filter.

Figure 4-15.

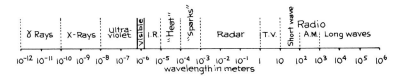

Figure 4-16

Average wavelengths of colors

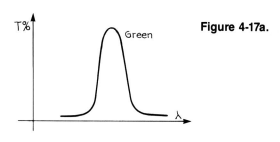

Figure 4-17a.

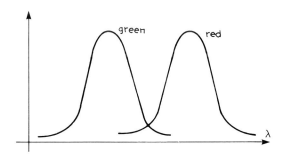

Figure 4-17.

IV.6.2. — THE ROLE OF COLOR FILTERS

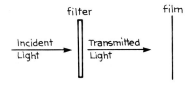

Figure 4-18.

Figure 4-18 shows a filter placed within the path of a beam of light; the light transmitted by the filter then arrives on the film. For each wavelength the intensity of the transmitted light is equal to the intensity of the incident light multiplied by the filter's coefficient of transmission for the wavelength under consideration.

But for most filters, the coefficient of transmission depends upon the wavelength. These are called "color filters" because they let certain colors pass through more easily than others.

As an example, Figure 4-17a presents the transmission curve of a green filter.

Figure 4-19.

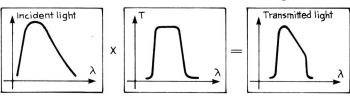

If we now multiply, wavelength by wavelength, the intensity of the light transmitted toward the film by the film's chromatic sensitivity, we obtain the contribution of each wavelength to the blackening of the film.

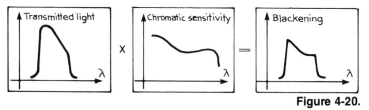

Figure 4-20.

By way of definition, we see that the blackening of the film is the result of multiplying the chromatic intensity of the source (incident light) by the product of the transmission of the filter and the chromatic sensitivity of the film. We shall refer to this product generally as the "chromatic (or spectral) response of the filter-film combination."

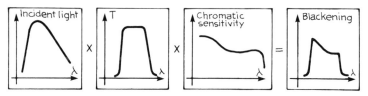

Figure 4-21.

In this presentation, the filter seems to us like the tool which allows us to modify the sensitivity curve of the film.

Figure 4-22.

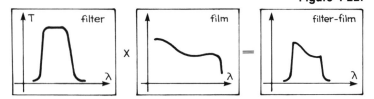

It is necessary, then, to state several particular cases—for example, the association of a red filter with an orthochromatic film.

Figure 4-23.

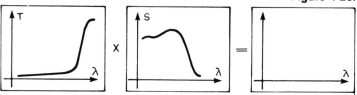

The result is . . . a uniformly nil sensitivity: no light can affect the film!

Note finally that you can use a combination of several filters to obtain the desired chromatic response. The curve of transmission for a group of filters is equal to the product of the transmission curves for each:

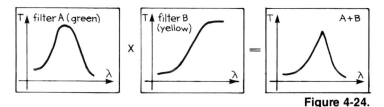

Figure 4-24.

IV.6.3. — WHY USE A FILTER?

We have seen that the contrast of the object photographed greatly influences the resolving power of films. Now, this contrast sometimes depends upon the wavelength.

Take for example the famous Red Spot of Jupiter. This spot appears within Jupiter's STB—South Tropical Belt (Figure 4-25).

If the spot appears red to us, the reason is not that it is redder than the STB, but that it is less blue. Consider the spectrum of light originating from the Red Spot and from the STB as shown in Figure 4-26.

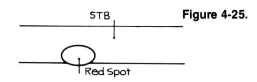

Figure 4-25.

Figure 4-26.

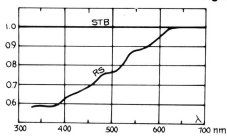

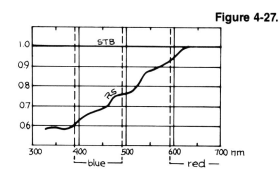

Figure 4-27.

IV.6.4. — FILTERS ADAPTED TO ASTROPHOTOGRAPHY

The astronomer seeks to select certain wavelengths while losing the least amount of light possible. For him the ideal filter is one which lets pass the maximum of light in the color selected and which becomes brutally opaque outside it. Its transmission curve resembles a crenel (Figure 4-28). Unfortunately, it is rare to find filters whose transmission curves are so straightforward.

These two curves epitomize the entire phenomenon. We see that the dominant color of the Spot is indeed red. On the other hand, we see that the total luminosity of the Spot (all wavelengths mixed) is less than that of the STB. Thus it appears to us like a dark point on a light background. As a result, these curves show us that the difference in luminosity between the Red Spot and the STB is greatest for blue light and very small for red light.

Thus, it is quite clear that the last thing to do to photograph the Red Spot is to use . . . a red filter. On the contrary, a blue filter lets you obtain a contrast superior to that of white light, and hence a greater efficiency (Figure 4-27).

We have just shown in this example the advantage provided by filters. You have to be conscious, however, of a major inconvenience: since every filtering process occurs through the absorption of a portion of the light, the use of a filter always results in a lowering of the total luminosity. Now, apart from the sun, we know that we are always coming up against the problem of too faint a luminosity.

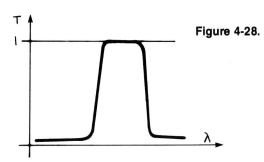

Figure 4-28.

Kodak markets under the name Wratten a series of very interesting, easily available filters. Most Wratten filters are of a flexible gelatin. They have the advantage of being less costly and can be cut into the desired shape to accommodate the desired site in the optical system. Figures 4-29, 30 and 31 shows the transmission curves for a few Wratten filters usable in astronomy.

The German firm Schott makes glass filters which complete the Wratten spectrum, especially for windows of transmission situated in the violet or ultraviolet. Here are a few examples of transmission curves for Schott filters (Figure 4-32 and 33).

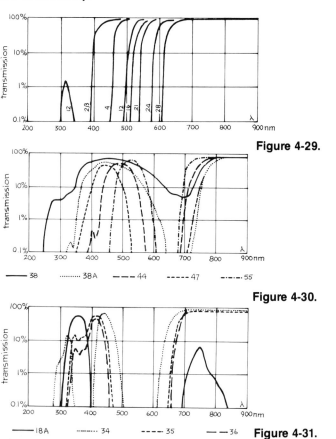

Figure 4-29.

— 38   ········ 38A   — — 44   — — — 47   —·—·— 55

Figure 4-30.

——18A   ········ 34   — — — 35   — — 36

Figure 4-31.

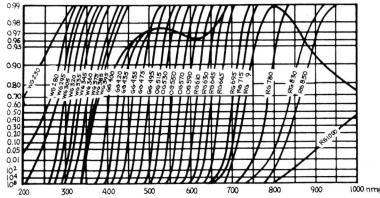

Figure 4-32.

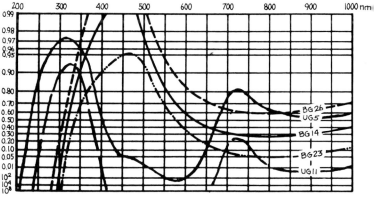

Figure 4-33.

The French firm MTO can supply on demand color filters of various frequency ranges as well as interference filters. The latter, of a very narrow frequency range, are used when you wish to isolate a particular ray in the spectrum of an object.

IV.7. — EQUATION OF LUMINOSITY

In this section we shall examine the parameters involved in the quantity of light received by the film. These parameters should be combined so that this quantity of light provides a correct exposure for the film. They should thus satisfy among themselves the relation that we are about to determine.

This relation is intended for the general case when we are considering an object presenting a large apparent surface with respect to the resolving power of the telescope. The other case—that of point objects—has been covered in the context of Section II.7.

IV.7.1. — FILM EXPOSURE—SPEED

We have seen with the characteristic curves of films that a photograph is correctly exposed when the blackening attains an intermediate value corresponding approximately to the maximal reproduction of contrast. In fact, considerations of the signal-to-noise ratio and the equivalent quantum efficiency place this optimum at an exposure slightly less than that of the straight-line portion of the characteristic curve (maximum contrast).

To this optimum blackening there corresponds a certain exposure, that is, a certain quantity of light received per unit of surface. This desired exposure is, moreover, inversely proportional to the speed of the film, by the very definition of this speed.

IV.7.2. — EXPOSURE TIME

It is quite clear that, everything else being equal, the quantity of light received by the film is proportional to the exposure time.

Here we shall not take into account reciprocity failure; thus, the relation established among the parameters will be valid for exposures not too much longer than a second, which is generally the case for planetary photography, the area for which this relation is the most useful.

For longer exposure times t we must simply introduce a corrective term. In the first approximation, we can consider that everything happens as if the quantity of light were no longer proportional to t, but to $t^{0.8}$ (Schwarzschild's approximation).

IV.7.3. — OBJECTIVE DIAMETER

Of course, apart from the losses due to absorption in optical systems (which we shall ignore here) the quantity of light which reaches the film is equal to the quantity of light which penetrates the telescope, that is, which crosses the objective (Figure 4-34). This quantity of light is thus proportional to the surface of the objective and hence to the square of its diameter (D^2).

Figure 4-34.

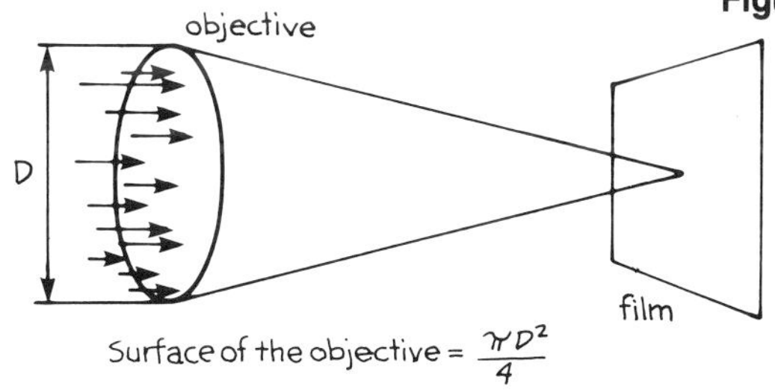

Surface of the objective $= \dfrac{\pi D^2}{4}$

IV.7.4. — EFFECTIVE FOCAL LENGTH

Imagine that we are taking a picture of Mars. All the light from this planet captured by the telescope is distributed on the film on a surface which represents the image of Mars. We know that the diameter of this image is proportional to the effective focal length used (see Section III.2.); thus, its surface is proportional to the square of this equivalent focal length (F^2).

Now, what is important for the exposure of the film is the density of the light, that is, the quantity of light received per unit of surface. This

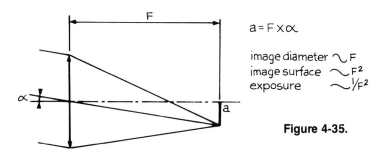

$a = F \times \alpha$

image diameter $\sim F$
image surface $\sim F^2$
exposure $\sim 1/F^2$

Figure 4-35.

density is inversely proportional to the surface on which the light is distributed, that is, the surface of the formed image. The density of the light (or exposure) is thus inversely proportional to the square of the effective focal length $(1/F^2)$. See Figure 4-35.

IV.7.5. — APPARENT LUMINOSITY OF THE OBJECT

It is clear that the quantity of light received by the telescope—and hence by the film—is proportional to the apparent luminosity L of the photographic object. But this is a case of luminosity per unit of surface. Indeed, at equal total luminosity, a more extended heavenly body will give, for the same effective focal length, a larger image and thus a smaller density of light at film level.

Appendix B gives a calculation for evaluating surface luminosities.

IV.7.6. — SYNTHESIZING THE INFLUENCE OF ALL THE PARAMETERS

We have seen that the blackening of a film is directly proportional to its speed S, to the exposure time t, to the surface luminosity L of the object, to the square of the diameter D of the telescope, and inversely proportional to the square of the effective focal length F of the system. Now, this blackening must fall within a well-determined region, independent of the conditions of shooting. Thus, we can write the following equation:

$$(L\ S\ t\ D^2)/F^2 = \text{constant},$$

which must always be satisfied in order to obtain a correct exposure of the film.

The actual value of the constant which this ratio must always equal is not important, since the amateur astronomer never uses it in the absolute, but always for comparison.

Let's take a few examples. For D = 100mm, F = 8000mm, S = 100 (ASA), the exposure necessary for Jupiter is 2 sec. If we use a focal length twice as long, or 16,000mm, the term F^2 is multiplied by 4 and we will need to use a film 4 times faster (ASA 400) or an exposure 4 times as long in order to keep our ratio constant. On the other hand, if we are photographing Mars, which has a surface luminosity 4 times brighter than Jupiter's, we will be able to make do with a film 4 times slower (ASA 25) for the same focal length (8000mm) and the same exposure time (2 sec.). And so on.

It is absolutely fundamental to know that there is a relation of interdependence among all the parameters of a shot (luminosity of the object, speed of the film, exposure time, diameter and focal length of the telescope) and to know the influence that each has upon the search for the correct exposure for the film.

From now on, we shall interest ourselves especially in the search for an optimal combination of the parameters for a given object. For this object, the luminosity L will thus remain constant in every case envisioned for the course of the discussion. We can thus write our equation more simply:

$$(S\ t\ D^2)/F^2 = \text{constant}.$$

Common practice is to regroup the terms F and D, which "dimension" the optical system used:

$$(S\ t) / (F/D^2 = \text{constant}.$$

Under this form the direct relation between the diameter and focal length of the telescope becomes more apparent: an objective twice as large permits us to use a focal length twice as long and hence obtain an image twice as large.

It is under this last form that terrestial photographers really know the luminosity equation. They attach to their camera the speed of the film used and then use the readings from a photoelectric cell to find, according to this speed S, what exposure time T corresponds to the F/D ratio (f-stop) chosen, or vice versa.

But here a difference arises between the astronomer and the photographer. The latter first chooses the focal length to use according to the field desired and then varies the f-stop merely by playing with useful diameters for the objective. Indeed, the objectives are generally of fixed focal length (zooms are not an exception, since they maintain a constant F/D ratio when F varies) and the f-stop is achieved with the help of a ring which works a diaphragm placed on the objective. In astronomy, it is out of the question to diaphragm the objective of the telescope, since we know that the fineness of the images is related to its diameter. Thus, a telescope is always used at its optimal aperture. On the other hand, the photographic object is generally small with respect to the available field, which lets the astonomer choose the F/D ratio he likes by varying the effective focal length of his setup.

If the astronomer uses a filter, he must take this into account in the luminosity equation. The easiest way to include this parameter is to consider that the available light is equal to the intensity of the incident light multiplied by the transmission factor T of the filter, which gives:

$$(T\,S\,t) / (F/D^2) = \text{constant.}$$

T ranges from 0 (perfectly opaque filter) to 1 (perfectly transparent filter or absence of filter).

This last equation poses no problem if the filter is a dimming filter whose transmission factor varies little with the wavelength. Now, the filters used in astronomy are rather selective and hence present a factor of transmission which varies from one part of the spectrum to another. In this case, we have seen (Section IV.6.2.) that the influence of the filter can depend upon the chromatic response of the film. It is thus more correct to consider in a related fashion the filter-film coupling and to write our equation:

$$(S_e\,t)/(F/D^2) = \text{constant,}$$

where S_e is the effective speed of the film used with a chosen filter.

This presentation may seem surprising, but it is well adapted to our discussion, since we consider the filter as a means of increasing the resolving power of the film (by bringing about an increase in the image contrast).

IV.7.7. — THE PROBLEM OF CHOICE

We have just examined a number of parameters which we should keep in mind when we take a photograph. These are, for a given object and telescope:

- the sensitivity of the filter-film-developer combination
- the F/D ratio
- the exposure time.

These three parameters are not independent, since they must satisfy the luminosity equation.

We thus have complete freedom in choosing two of them, but the third will be imposed upon us by this choice in order to satisfy the equation.

In spite of this restriction, the possible choices are particularly vast. It is up to us to determine the optimal combination, that is, the one which lets us exploit to the maximum the performance ability of the telescope used.

Unfortunately, we discover that every choice whose goal is to increase the resolution of the photograph results in rendering the system less capable of capturing faint sources of light, and vice versa. In fact:

In general, the faster the film, the lower its resolving power. In the same way, an active developer which lets us "push" the speed of the film always degrades its resolution. Conversely, a filter—which would permit an increased performance in the film—causes a loss of light.

A higher F/D ratio, which lets you obtain a sufficiently enlarged image of the photographic object, will cost you, from the point of view of the luminosity equation, by the dispersion of light it causes.

Finally, a long exposure time, which lets you compensate for the use of a large F/D ratio and a slow film, makes the photograph sensitive to three parasitic phenomena: tracking errors, vibrations, and atmospheric turbulence.

Thus, you will notice at this stage of the discussion that the optimum choice of parameters must represent a compromise between the necessity to expose the film sufficiently and the desire to use at the same time a powerful film, a large image, and a short exposure time.

Clearly the best solution will depend upon the object that you want to photograph. In particular, the results will be very different according to whether you are photographing the sun, planets, or faint objects. We are thus obliged from now on to treat these three cases separately.

We shall begin by treating planetary photography, for which the choice of parameters is the most critical.

CHAPTER V
PLANETARY PHOTOGRAPHY

You will recall that planetary objects consist of the brilliant planets (Mercury, Venus, Mars, Jupiter, Saturn) and the moon. These objects are characterized by an apparent surface discernible in amateur instruments and by a moderate luminosity. This luminosity is sufficient for making high-resolution photos—but not without difficulties. High-resolution photography consists of obtaining the smallest visible detail and, if possible, of attaining the theoretical resolving power of the instrument.

To reach this end, you must choose the parameters of your photography properly and devote the utmost care in their execution.

V.1. — CHOICE OF FILM, DEVELOPER, AND FILTER

We can not separate these three parameters. In fact, they are characterized together by image contrast, sensitivity, and resolution.

V.1.1. — CONTRAST

In general, the higher the contrast of the negative, the more the image reveals fine details. In astronomy, where planetary surfaces tend to be rather uniform, we seek to obtain images as high in contrast as possible.

What is the upper contrast limit? It depends upon the object photographed and what you want to see. In general, we are limited by the fact that two different areas of the object have very different luminosities. A high-contrast film can thus give an overexposed zone and an underexposed zone for the same object, thus limiting the useable portion. For example, we are interested in using the highest contrast possible to discern the bands and spots of Jupiter, but in so doing we lose the edge of the planet, which is darker than the center. With the moon on a contrasty film, it is impossible to obtain all the details of the terminator without overexposing the illuminated portion. We run into the same problem with Saturn, the ring being darker than the planet.

We shall adopt the following rule: the contrast of the film should be the highest possible so that, of the areas of interest on the object photographed, the darkest and brightest areas are just at the limits of underexposure and saturation, respectively.

How do we obtain the desired contrast? First of all by the choice of film, then by the choice of developer and the method of development. We have seen that the characteristic curve (whose slope is nothing but the contrast) varies from one film to another. Moreover, the more energetic the developer is, and the more prolonged the developing bath, the more contrast there will be in the negative.

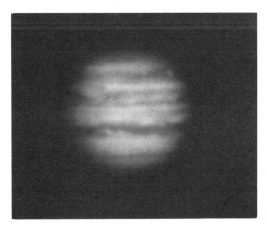

Photograph 5-1. *Jupiter photographed by Jean Dragesco using a 355mm telescope with a F/D = 50 on Kodak 2415 film exposed for 2 seconds.*

V.1.2. — THE RESOLUTION—SPEED RATIO

Let's now set aside the film's contrast, or rather, suppose it to be adapted to give a discernible picture over the entire field, all with the best resolution of fine details.

We know that we need the fastest possible film with the highest possible resolving power. But, these two qualities are contradictory: as a general rule, the faster the film, the less resolution it gives. The same holds true even when you develop a given film with an energetic developer for a prolonged time.

So, should we choose a high-resolution film or a very fast film which will allow us to use a longer focal length? To answer this question, let's take as an example the comparison of two films, one having a speed of ASA 100, the other a speed of ASA 400 and hence four times as sensitive. We must compare these two emulsions under the same conditions. In particular, the exposure time—which intervenes in problems of turbulence and blurring—should be the same. The focal length used for the ASA 400 film is thus twice that allowed by the ASA 100 film (see the luminosity equation introduced in the preceding chapter, relating speed, focal length, and exposure time). As a result, the image obtained on the negative is twice as large for the ASA 400 film. Therefore, it requires only a spatial resolution two times less fine than that of the ASA 100 film in order to reveal the same details.

If under certain circumstances of exposure the ASA 100 film resolves details with a diameter of $20 \mu m$, the ASA 400 film, in order to be equivalent, must resolve $40 \mu m$ under the same circumstances. Now, our experience with films shows that the ASA 400 film will actually give us (in this example) a resolution of $30 \mu m$. For high-resolution planetary photos, when there is no problem with field, it thus seems preferable to use—at least with ordinary films—fast films developed beyond their nominal speeds with an energetic developer, together with long focal lengths.

Apart from the speed-resolution dilemma, this solution has two advantages: the energetic developer gives a higher image contrast (to which we must often attribute the gain in resolution anyway) and the formation of a larger image can facilitate focusing.

V.1.3. — THE CHOICE OF FILTER

A colored filter, selective for certain wavelengths, can improve image quality for three reasons: heightened contrast of planetary details, elimination of atmospheric diffusion, and reduction of the effects of chromatic aberration.

A heightened image contrast can be obtained by selecting wavelengths for which the difference in luminous density between neighboring planetary details is the largest. Therefore a blue filter enhances the Red Spot of Jupiter, which is darker and less blue than the zones surrounding it. But the contribution made by filters can not be generalized. On the moon, for instance, there is not a significant difference in coloration between the edge and the bottom of a crater—thus, no filter can improve the contrast at this level.

Depending upon its purity, the earth's atmosphere more or less diffuses the light passing through it. This phenomenon is especially annoying when observing the moon, since it is the most extended of the celestial bodies, or indeed when observing another object close to the moon. Atmospheric diffusion can be likened to the superimposition of a luminous fog upon the original image, whose contrast is thus diminished. Since this diffusion is more pronounced for short wavelengths, the best remedy consists of eliminating the violet and blue light by using a yellow filter such as the Wratten 8 or the Wratten 12.

Finally, we know that an image can present appreciable chromatic aberration (see Section I.4.8.) which is instrumental in origin if the telescope is a refractor and atmospheric in origin if the object is far from the zenith. The simplest solution consists of reducing the spectral range of the incident light by using a selective filter. The optics of refractors generally present a minimum of residual chromatic aberration in the green and yellow; thus, it would be profitable to eliminate the violet and the blue on one side, the orange and the red on the other. The "green" filters making this selection, however, have relatively small transmission factors. For this reason, we

Photograph 5-2. *First quarter moon photographed by Christian Arsidi using a 200mm telescope with a F/D = 10 on Agfapan 25 for 1/2 second.*

generally prefer instead a less severe "yellow" filter, which is opaque only for short wavelengths (see, for example, the curves of the Wratten 12 [yellow] and 55 [green] in Section IV.6.4.).

Note that atmospheric aberration can be countered without using a filter by placing in front of the telescope objective a glass prism whose chromatic dispersion compensates for that of the atmosphere. The optic quality necessary for such a prism, however, makes it very expensive. Furthermore, since atmospheric aberration depends upon the observed object's distance from the zenith, you would need an entire series of different prisms. This solution is thus not realistic for an amateur.

Filters, however, have one important inconvenience: they lose an enormous amount of light, which we translate as a reduction in the sensitivity of the film-developer-filter trio. Thus, we should use filters only when they permit us to bring out details whose contrast in white light is very low or nil (for example, the dark spots in the Venusian atmosphere, which are visible only in ultraviolet) or when the object's brightness (moon, Jupiter) permits the use of a yellow filter. In the other cases, the loss of light proves more costly than the chromatic aberration or lack of contrast.

V.1.4. — RECOMMENDED FILMS

Until recently, amateur astronomers had only ordinary commercial stock at their disposal. Outside of a few films with a sensitivity a little too weak for planetary photography, these emulsions all have a similar contrast. This contrast, well adapted to everyday photography, is of the order 0.7, that is, the image contrast furnished by the film is 70% of the contrast of the original subject. Planetary photography, on the contrary, demands an amplification in contrast of the original subject, which is naturally too low. Thus, we must subject these ordinary films to a more energetic development than their standard treatment (strong developer; prolonged bath time).

A series of tests designed to determine the resolving power of each of the common emulsions has shown the superior quality, in absolute value, of the slowest films. According to the criterion we have just defined, however—integrating the problems linked to speed and resolution—the fastest films seem on the whole better adapted to high-resolution astrophotography.

The same observation holds for developers: the gain in speed obtained by using an energetic developer compensates greatly in most cases for the loss in resolving power.

In the course of these tests, the film which gave the best results with respect to sensitivity is Kodak's Tri X developed in a very energetic developer (Kodak's D 19 or Paterson's Acuspeed).

Recently two films of especial interest to high-resolution planetary photography have become available. The first is Kodak's Technical Pan 2415. This emulsion provides a speed-resolution compromise superior to everything which had been known before, despite a somewhat slow speed (between ASA 25 and 200, depending upon the development). The other strong point in 2415's favor is its ability to offer a very heightened contrast. Thus, in the majority of cases, it supplants the ordinary films still used just a short time ago in planetary photography.

The other newcomer is variable-speed film, Agfa's Vario XL or Ilford's XP1. The latter, moreover, seems better adapted to astronomy since it can attain a higher speed with a similar resolution. The advantage of these "variable-speed" films is that they reduce the picture's macro-contrast because of their variable speed. The moon will thus present a correctly exposed terminator and illuminated zone at the same time. On the other side of the coin, the micro-contrast (that is, the contrast between two neighboring points in the picture) suffers a little, which impedes the search for fine details. XP1 is thus better adapted to lunar photography (where the details are generally of sufficient contrast and where the illumination can vary greatly from zone to zone) than for planetary photography, where the details sought are always low in contrast.

We saw in the preceding chapter that this resolution-speed criterion can be expressed rigorously by looking for a film whose equivalent quantum efficiency is the highest. Unfortunately, the measurements of this EQE have currently been done only by professionals for their own needs. No publication of EQE's for all the common films traditionally used by amateurs yet exists. Nevertheless, we can make the following classification according to decreasing EQE: Kodak 2415, Ilford XP1, Agfa Vario XL,

Photograph 5-3. The walled plane Plato and the bay Sinus Iridum on the moon photographed by Jean Dragesco using a 355mm telescope with a F/D = 40 on XP1 film exposed for 1 second.

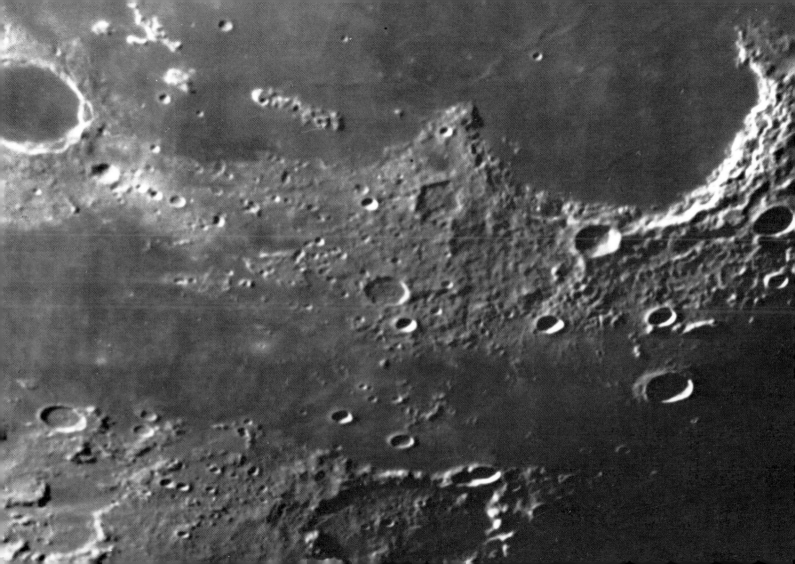

Tri X and other ordinary ASA 400 films. Thus, Kodak 2415 represents the optimum, but with a higher resolution and lower speed than the immediate competition. Hence, it will be used with a shorter focal length and thus a smaller image, so it is necessary to make sure that the optical system and especially the focusing—a more tedious operation with a small image—does not degrade the quality. If this is the case, a faster film will suit you better.

In color films, the best definition is obtained by Kodachrome 25 and Kodachrome 64. Like the black-and-white emulsions, however, the faster films are preferable. Ektachrome 200 seems to provide the best speed-resolution compromise.

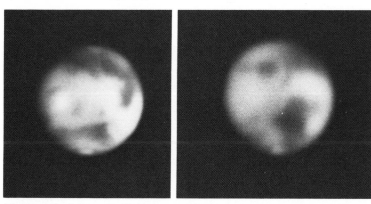

Photographs 5-4 a and b. These two photos of Mars were taken by Georges Viscardy on Kodak 2415 with a 520mm telescope with an effective focal length of 36 meters exposed for 1 second.

V.2. — CHOICE OF EXPOSURE TIME

The exposure time is limited by the untimely displacement of the image forming on the film. Obviously, every little movement depreciates the quality of the photography, and this movement becomes more substantial the longer the exposure time.

We can classify the causes of such image displacement into two categories: that supplied by the instrument and that due to the atmosphere.

The relative influence of these two categories varies greatly according to the instrument used. In fact, the small amateur telescopes often have mountings that are too rudimentary, generating vibrations and not permitting good tracking. On the other hand, the large telescopes, and in particular the professional instruments, are furnished with much more trustworthy motorized mountings. Nevertheless, their large diameters make them more sensitive to the spreading of images caused by atmospheric turbulence.

V.2.1. — VIBRATIONS

Poorly stabilized mountings, the tug on a shutter release cable, the movement of the camera's reflex mirror, or the slightest breath of wind become sources of vibrations, which die very slowly.

A vibration is a periodic phenomenon. Only a short exposure time during the period of vibration can minimize its effects. On the other hand, as you can see in Figure 5-1, the amount of blurring, which is defined by the amplitude of vibration, remains constant whenever the exposure time exceeds the period of vibration.

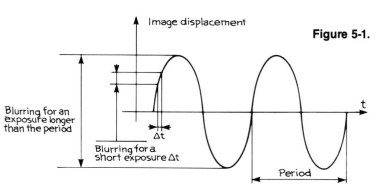

Figure 5-1.

Vibrations occurring with telescopes generally have periods of less than a second. It is not possible in planetary photography to use much shorter exposure times. Thus, we can not limit the detrimental effects of vibration with the exposure time.

Thus, the problem of vibrations must be treated at the source: do away with problem parts like cables; lift the reflex mirror manually several seconds before shooting, if you can do this with your camera; finally, reinforce mounting and support to make them as rigid as possible.

An excellent remedy for vibrations is a manual shutter. Place a mask in front of the telescope entrance before setting the camera in open position (T or B setting). Then, remove the mask by hand, taking care not to touch the tube, for the time necessary to expose the film. This method eliminates all vibrations originating with the camera mechanism, but it applies only to exposures greater than half a second.

V.2.2. — TRACKING ERRORS

V.2.2.1. — THE GENERAL CASE

Recall that in order to compensate for the apparent motion of the sky, the telescope must be turned with and East-to-West rotation, around an axis parallel to the axis of the terrestrial poles, at the speed of one turn in approximately 24 hours. This is the principle of the equatorial mounting.

If the telescope does not turn, the apparent motion of the sky will cause a blur on the photograph. The amount of blurring is proportional to the exposure time, since the sky travels at a constant speed.

The bodies we are interested in (moon, planets) are always close to the celestial equator. It is precisely in this region that the angular displacement of the sky is the most appreciable—15″ per second of time (the stars close to the pole hardly move during the night). If we decide upon 1″ as the goal to attain, which is a reasonable order of magnitude for an amateur scope, our exposure time will have to be shorter than 1/15 second if the telescope is not driven. Such a short exposure is too short for planetary photography. We must thus have an instrument equipped with a motorized equatorial mounting if we hope to have good results with high-resolution planetary photography.

If such is the case, however, then we must be conscious of two causes of bad tracking: on the one hand, the polar axis of the equatorial mounting might not be pointed exactly at the celestial pole, and on the other hand, the speed of rotation furnished by the motor and the drive system might be imprecise.

When the equatorial mounting's polar axis is poorly aligned, the most unfavorable case will occur when you aim at a body in a direction perpendicular to the plane defined by the telescope's axis of rotation and that of the earth. For example, if the axis of the telescope lies well within the proper meridian, but makes too large or too small an angle with horizontal, the directions where the image displacement is most appreciable are East and West.

In this case, the point followed by the telescope traces in the sky a line which makes with the path actually followed by the object an angle equal to the error in alignment of the polar axis (Figure 5-2).

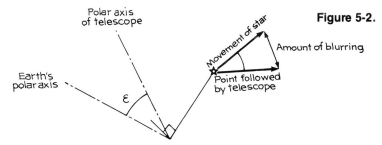

Figure 5-2.

We can approximate the amount of blurring with a simple formula:

blurring = alignment error in ° × displacement of sky/60.

Let's take an example. If the exposure time is 4 seconds, the apparent displacement of the sky at the equator is 4 × 15 = 60″. If the error in aiming the polar axis of the scope is 1°, then the image suffers a blurring of 1″, which is the maximum permissible value.

You can see the importance of precision in aligning the polar axis—in round figures, you must not allow an error greater than 1° for an exposure on the order of one second. This precision is easy to attain with a little practice, even on a portable instrument. If the telescope is permanently installed, the precision can be even greater (for a discussion on alignment, see Section VIII.1).

The other cause of defective tracking has to do with the drive speed. Depending upon the technology employed, the speed of the motor can be more or less precise. In addition, eccentricities in the drive system gearing or irregularities in the size of the teeth cause periodic accelerations and decelerations. If the overall imprecision of the rotation speed is 1%, the image movement is one hundredth of the displacement of the sky during the exposure. It attains one second of arc when the sky is displaced 100″, that is, at the end of about 7 seconds.

To sum up, admissible exposure times are proportional to the smallest angle you want to resolve on the film (we can tolerate an image displacement ten times greater if we accept a definition of 10″ and not 1″) and inversely proportional to the error committed by aiming the polar axis and by the regulation of the drive speed. Let's retain the following simple values guaranteeing a necessary margin of a factor of 3 to 4 between the image displacement and the desired resolving power: for an alignment error of 1° or for an imprecision of 2%, adopt an exposure, in seconds, equal to the desired resolving power expressed in seconds of arc.

The Table 5-1 gives the maximum exposure time called for by the desired resolving power and the tracking errors.

TABLE 5-1
Maximum Exposure Time (Sec) for Selected
Errors in Polar Alignment and Drive Speed

Error		Resolving Power			
Polar Alignment	Drive Speed	0.5″	1″	2″	4″
0.5°	1%	1	2	4	8
1.0°	2%	½	1	2	4
2.0°	4%	¼	½	1	2
4.0°	8%	⅛	¼	½	1

V.2.2.2. — LUNAR MOTION

The moon exhibits the peculiarity of having a substantial proper motion. Completing its orbit in 27 days and 7 hours, the moon has an average apparent motion in the sky of not 15″, but 14.45″ per second of time. If the equatorial drive is perfectly regulated for *stellar motion*, tracking the moon will cause an image displacement of one half second of arc per second of time. This is often not acceptable for high-resolution photography; thus, we must also be able to regulate the drive speed for lunar motion.

But our problems do not stop there. In fact, the lunar orbit is inclined 5° to the earth's orbit—but since the earth is itself set 23.5° in its orbit, the axis of revolution of the lunar orbit makes an angle of 28.5° with the earth's polar axis. Consequently, the moon's proper motion has a component on the axis of declination which can attain a maximum of 0.26″ per second of time. On certain days the movement in declination is by itself sufficient to prevent exposures longer than a second of time, if the desired resolution is one second of arc. To be perfectly precise, we would have to equip the equatorial mounting's axis of declination with a variable-speed motor drive. Amateurs rarely add this option because it seriously complicates installation and is really necessary only for long exposures when the moon's declination lies in the neighborhood of zero (moment of the most rapid variation in declination).

V.2.3. — TURBULENCE

V.2.3.1. — WHAT IS TURBULENCE?

Like all transparent media, air has an index of refraction different from 1; thus, it is capable of bending the light rays crossing it. Now, the index of refraction for air depends upon its temperature. Every light ray crossing layers of air of different temperatures undergoes deviation at each border between these layers, since it is passing into media with different indices (Figure 5-3).

Photograph 5-5. The region of Lacus Somniorum on the moon photographed by Jacques Barthes with a 210mm telescope on Ilford FP4 film for 10 seconds and developed in Microphen. This rather long exposure was facilitated by very stable fork mounting.

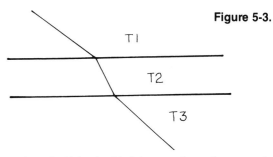

Figure 5-3.

Now imagine a bubble of cold air in a medium of warmer air. The index for the bubble is higher—it thus plays the role of a converging lens (Figure 5-4). The light rays which were parallel before crossing the bubble (because they originate from the same star, for example) afterwards follow slightly different paths. For the observer, these rays of light no longer seem to come from a single point, but from a multitude of points more or less dispersed.

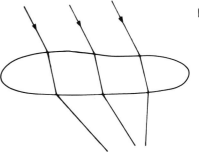

Figure 5-4.

But bubbles of air such as we have just described are not stable—they deform and displace themselves quickly, so that the light rays do not remain bent in a constant manner. The observer thus sees a multitude of luminous points rapidly dancing. The stars, theoretically points, seem like little spots with fleeting tufts, while the limbs of the moon and sun seem to bubble. As for planetary details, they seem more or less fuzzy and very often appear to move with respect to one another.

In practice, we can consider the detrimental effects of turbulence as being the sum of two phenomena:

1) The spreading out of images, generally related to turbulence at low altitudes or at the instrument level. Every normal point source takes on a thickened, planetary aspect. Since the light rays which have been dispersed reach the observer from different directions at the same time, it is hard to counteract the thickening by a reduction in exposure time.

2) The agitation of images, generally related to turbulence at high altitudes. The light sources (stars, planetary details, etc.) are not thickened, but they tend to be displaced erratically. This kind of displacement implies a relation between the exposure time and the resolution of the photograph.

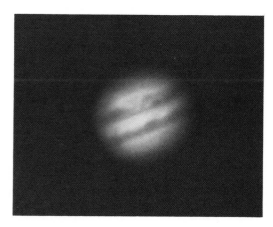

Photograph 5-6. *Christian Arsidi took this picture of Jupiter using a 280mm telescope with an effective focal length of 16 meters on Kodak 2415 film exposed for 2 seconds.*

V.2.3.2. — THE CAUSES OF TURBULENCE

Image alteration occurs whenever light rays cross a mass of air which is not homogeneous in movement. The principal sources of turbulence can originate from:

1) The general condition of the atmosphere. The wind and strong temperature gradients are important factors in turbulence. Conversely, a calm atmosphere with (in the winter) just a few light banks of mist generally permits good images. The turbulence related to high layers of the atmosphere is generally characterized by erratic displacement of stellar images and planetary details about their average position, rather than by a blurring of the image.

2) The observation site. The altitude of the site is a very important factor. When the observatory is high, the rays of light do not have to pass through the lower layers of the atmosphere, which are the densest and most disruptive. Furthermore, the environment of the location of the site has an appreciable influence. There is generally more turbulence at the bottom of a valley than at the top of a hill. In the evening foliage has a beneficial effect by radiating little heat into the ambient air. Of course, cities are prohibitive: in summer because of the heat stored up during the day by concrete and tarred surfaces, which is returned in the evening; in winter because urban centers (owing to their innumerable chimneys) generate immense columns of warm air mixing with the colder ambient air to create enormous turbulence. This type of turbulence, localized in the bottom layers of the atmosphere, produces widened, deformed stellar images that seem to pulsate. Planetary details seem to boil from time to time.

3) The instrument itself. Heat transfer between the walls of the tube and the air sometimes transform the tube into a chimney. To convince yourself of this, try observing on

a cold winter evening with a reflector you have just taken out of a warm apartment. The tube reheats the layers of air close by. This warm air rises and escapes from the opening of the tube, replaced by the cold air which engulfs it. This turbulence can attain enormous proportions (several dozen seconds of arc). The remedy consists of using, if possible, an instrument with a closed tube (refractor, or reflector equipped with a optical window) and in all cases taking care that the instrument is stabilized to the ambient temperature at the moment of observation. We shall see that turbulence due to the tube is a much thornier problem in solar photography, since the sun's light is a supplementary source of heat at the telescope focus.

V.2.3.3. — THE INFLUENCE OF TELESCOPE DIAMETER

The larger a telescope's diameter, the greater the resolving power will be. Large telescope users, therefore, must be more demanding about turbulence and hence, among other things, about the quality of the site. For instance, a turbulence 1″ of arc in amplitude passes practically unobserved in a 60mm refractor (resolving power = 2″) but proves catastrophic for a 1m reflector (theoretical resolving power = 0.12″).

Furthermore, the volume itself increases with the diameter. In fact, the light rays gathered by a telescope cross the atmosphere as a cylinder whose base is the surface of the objective (Figure 5-5). The larger the objective, the larger this cylinder will be and the greater the chance that the light rays will pass through heterogeneous zones and thus be deviated differently.

V.2.3.4. — TURBULENCE AND EXPOSURE TIME

Observers must be aware of the problems of turbulence and do their utmost to limit its impact through a judicious choice of site and through particular care given to the thermal condition of the telescope.

We can never eliminate turbulence altogether, however, so we must reduce its effects upon photography by imposing a constraint upon the exposure time.

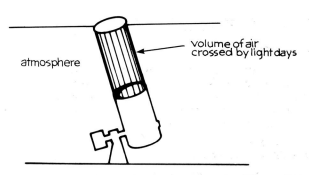

atmosphere

volume of air crossed by light days

Figure 5-5.

The light-ray deviations of the greatest amplitude usually have a rather long period (a few seconds). Visually these deviations are conveyed by small movements of the image rather than by a loss of sharpness. Thus, the exposure time must be relatively short. In this case a second of time is the maximum admissible order to magnitude for the exposure.

Turbulence, moreover, is not a constant phenomenon. By watching the image, attentive operators notice that from time to time there appear brief moments of calm—and it's up to them to snap the photograph at the right moment. But here as well, we can consider a second to be the maximal duration we can usually hope for within a lull.

Some amateurs, using cameras permitting multiple exposures, prefer instead of a single exposure of 1 second, four consecutive exposures of ¼ sec.; each of which, of course are taken on the same negative and in moments of calm.

Appendix C gives a method for evaluating the amplitude of atmospheric turbulence.

V.3. — CHOICE OF FOCAL LENGTH

V.3.1. — GENERAL CASE

For photographing a planet or a small region of the moon, there are generally no problems with the field covered by the dimension of the negative, and thus no additional constraint upon the focal length. For example, Jupiter, the largest of the planets, fills a 35mm format only when the focal length exceeds 100m.

The focal length of the optical system used must therefore furnish only an F/D ratio compatible with the luminosity equation.

Let us emphasize here that the logical course for determining the best parameters of a high-resolution photograph consists of choosing in order:

—the film and its developer, occasionally the filter, according to the necessary contrast and the best speed-resolution ratio;

—the exposure time, according to the performance of the motorized equatorial mounting, and the conditions of turbulence;

—finally, the focal ratio, imposed for a given object by the exposure necessary to register the film (luminosity equation):

(speed × exposure time)/(F/D)² = constant;

this constant, of course, depends upon the object photographed.

Unfortunately, amateur astronomers often try to follow a less logical course, first deciding upon the focal length and then the film; thus, the exposure time is dictated by the luminosity equation.

However, you must make sure that the determined focal length will guarantee that the smallest details you want to record are of a size compatible with the resolving power of the film. If this were not the case, it is certainly not the film you must change, since it already represents an optimum speed-resolution. You must therefore increase the exposure time slightly, which of course permits a longer focal length, until the angular resolution is as limited by turbulence (or the tracking) as by the size of the image on the film.

The following diagram will help us to understand this search for the optimum a little better. The film and developer having been determined, the dimension a_0 of the smallest detail visible in the image and its angular diameter α_0 are related by the focal length F: $a_0 = F \times \alpha_0$. Now, a_0 depends only upon the film; thus it is a constant. We can therefore say that the smallest angle α_0 visible on the photograph is—from the sole point of view of the limitation involved in the film's resolving power a_0—inversely proportional to the focal length: $\alpha_0 = a_0/F$. For its part, the exposure time t leads to a minimal angular resolution α_1 that depends upon the turbulence or the drift in tracking. In these two cases, if t is close to a second, which is the general case in planetary photography, we can allow that α_1 is proportional to t, thus to F^2 (since t and F^2 are proportional in the luminosity equation). The smallest angular diameter α visible on the photo is limited by the sum of α_0 and α_1. Figure 5-6 traces this limit as a function of the focal length. The curves show us that there exists an optimal focal length F_0 for which the smallest details will be visible. This optimum F_0 depends upon the relative sizes of the curves α_0 and α_1, and thus upon the film used, but also upon the reliability of the mounting and the conditions of turbulence!

Figure 5-6.

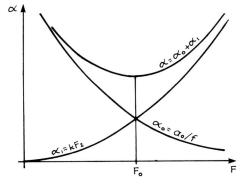

It is a good idea to remember the following law: an angular diameter equal to the resolving power of a telescope of diameter D is represented on the film by an image of dimension proportional to the ratio F/D. The dimension of this image is 0.1mm when F/D = 170. Knowing the resolving power of the film used, we can deduce from it the minimum F/D to have a chance of attaining the instrument's resolving power. For example, with Tri X, we can allow 0.03mm as the resolving power. The minimal F/D is thus $170 \times 0.03/0.1 = 51$.

V.3.2. — THE MOON IN ITS ENTIRETY

In planetary photography the only time the amateur astronomer need be concerned with the photographic field is when dealing with the full moon. The diameter of the moon can attain 33' of arc. In order to contain the full moon completely in 35mm format, it must have a diameter on the film of less than 24mm, which requires a focal length of less than 2.5m. Moreover, we advise you to limit yourself to 2.3m to avoid centering problems.

This focal length is almost always less than the optimal focal length; thus, you must expect not to have a photograph of optimum quality. You can, however, reduce this handicap by choosing a finer film. Be careful, though, not to choose too slow a film: the result would prove more detrimental because of increased exposure time.

V.4. — SUGGESTED SOLUTIONS

We are now going to apply the theory which we have just developed to the various subjects of high-resolution planetary photography.

V.4.1. — THE MOON

Because of the differing illuminations of lunar regions as their distance from the terminator increases, we do not recommend too contrasty a film-developer combination. The finest resolutions are to be sought near the terminator. On the one hand, the variations in lighting there are very great; on the other hand, the small details of lunar relief are highlighted by long shadows, thus creating contrasty images without the use of a high-contrast film.

Photograph 5-7. Here Jean-Paul Trachier recorded the passage of Io across Jupiter using a 310mm telescope with a F/D = 7 on Kodak Microfilm exposed for 2 seconds.

Thus, you should look for a film whose contrast lies between 0.6 and 1, which is the contrast range for standard films.

Kodak 2415 provides the best speed-resolution compromise, but in order to obtain the desired contrast, you must develop it in Pota, in Technidol LC, or in low-concentration HC110 (dilution F).

However, Kodak 2415 developed in this way loses much of its sensitivity—its performance thus becomes comparable to that of Ilford XP1,

with less resolving power but clearly more sensitive. The latter, moreover, presents a naturally weak contrast which makes it less suitable for planetary photography, but well suited for lunar photography. One of XP1's inconveniences is its processing, which resembles that of color films, and is thus rather delicate, requiring a rigorous bath temperature of 38 °C.

With its high definition but low sensitivity, Kodak 2415 requires short focal ratios, whereas Ilford XP1, lower in resolving power but more sensitive, needs longer focal ratios. The choice between these two films will depend upon the possible exposure times: you should prefer 2415 if you want a large field, XP1 if the telescope (refractor or Cassegrain) does not easily permit a short focal length.

With the full moon there is no terminator, the surface illumination is therefore rather "flat" and lacks high contrast. You must thus choose a film-developer combination which is higher in contrast than for the other phases. Kodak 2415 developed in D76 or HC110 (Dilution B or D) works very well.

Since the moon does not have much variation in color from one point to another, the use of a filter is not justified and color offer no advantage over black-and-white films.

The choice of exposure time depends upon the telescope mounting and atmospheric turbulence. These considerations have been developed for the general case (Section V.2). The F/D ratio is thus related to the speed of the film and to the exposure time by the luminosity equation (Section V.3).

Figure 5-8 gives the correct F/D ratio to use for different exposures t as a function of the moon's phase assuming a film of ASA100. The phase of the moon is given as the fraction of the disc illuminated (see side sketch). The graph indicates the days of lunation corresponding approximately to the different values for the fraction illuminated. The latter, a more exact characteristic than the phase, is published in astronomical ephemerides.

Photograph 5-8. The moon three days after first quarter photographed by Christian Viladrich using a 200mm telescope with an effective focal length of 2 meters on Kodak Microfilm exposed for 1/2 second.

For a photo of the moon in its entirety, the optimal focal length to use has a value set at about 2.4m, which requires a given focal ratio for each instrument. In this case, then, we use a graph giving the exposure time for various ratios as a function of the moon's phase.

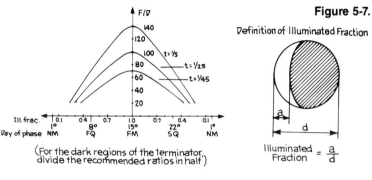

Figure 5-7

Definition of Illuminated Fraction

$$\frac{\text{Illuminated}}{\text{Fraction}} = \frac{a}{d}$$

(For the dark regions of the terminator, divide the recommended ratios in half)

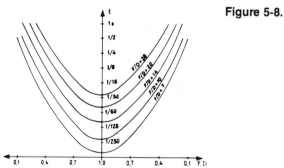

Figure 5-8.

Please keep in mind that the values furnished by these graphs are average values. They are to be slightly corrected according to whether you are interested in the terminator or the limb. Also, the loss of light in the optical system and especially the variable transparence of the sky can lead the operator to introduce corrections of a factor of 2 for exposure times and thus a factor of 1.4 for focal ratios.

Photograph 5-10. *The lunar crater Copernicus photographed by Christian Arsidi using a 280mm telescope with an effective focal length of 22 meters on Ilford XP1 film exposed for 1 second.*

Photograph 5-9. *The full moon (opposite page) photographed by the Author using a 90mm refractor with an effective focal length of 2.4 meters on Kodak 2415 film for 1/15 second and developed in D76.*

Photograph 5-11. *The lunar craters Clavius and Tycho photographed by Jean Dragesco using a 355mm telescope with F/D = 40 on Ilford XP1 film exposed for 1 second.*

V.4.2. — MARS, JUPITER, SATURN

These are the three planets most accessible to the amateur astronomer. We treat them simultaneously because their photography has many common points.

The contrasts among planetary details are rather low—you must use a high-contrast film. Kodak 2415 developed in D19 or Acuspeed (speed $\cong$ ASA 125; γ = 2.9) provides the best solution, from the point of view of contrast as well as speed-resolution compromise.

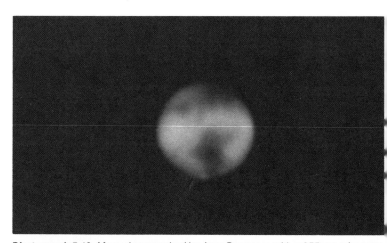

Photograph 5-16. *Mars photographed by Jean Dragesco with a 355mm telescope for an F/D = 60 on Kodak 2415 film exposed for 2 seconds.*

Photographs 5-15 a, b, c. *These three pictures of Mars were taken by Christian Viladrich on April 7, 1982, at 21:07, 22:28, and 22:55 U.T. with a 200mm telescope on Kodak SO115 film (now 2415) exposed for 1/2 second (a and b) and 1 second (c). The third exposure was taken with a Wratten 12 (orange) filter to augment the contrast of the dark spots. This series of photographs shows the planet's rotation.*

Jupiter's edge is always less luminous than its center. With a film like 2415 developed in D19, it is hard during printing to bring out the details of the center without obscuring the edges. The same problem occurs with Saturn, whose ring is clearly less luminous than the planet. You must not, however, give in to the temptation of a lower-contrast film (Tri X + D19, for instance) because the planetary details will then become less apparent. The center-edge effect must be attenuated in the darkroom with appropriate masking (see Section VIII.4.2).

Some filters allow you to augment the contrast of planetary details. With Jupiter, a blue filter (Wratten 38 or 38A) permits a better view of the Red Spot, whereas a yellow filter lets you better distinguish between certain bands. Mars can be photographed with an orange filter which accentuates the darkening of certain regions. All these filters were very popular among planetary photographers a few years ago, when a high-contrast film of sufficient speed did not exist. The appearance of 2415 has allowed us to capture low contrast details in white light and thus to dispense with the use of filters and the harmful loss of light that they cause. Except for particular cases (for example, if you want to better single out the Red Spot of Jupiter at the expense of other details) the use of filters with Kodak 2415 does not bring an important amelioration in image quality.

As far as color films go, the best speed-resolution ratios suggest fast films: Fujichrome 400, and especially Ektachrome 800/1600 developed at ASA 1600.

The surface luminosity of Mars is around four times greater than that of Jupiter, which in turn is nearly three times greater than that of Saturn. Everything else being equal, the exposure times for these planets thus differ by a factor of 3 to 4.

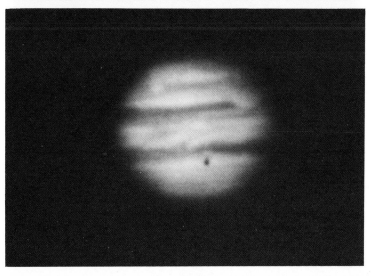

Photograph 5-12. *Jupiter photographed by Georges Viscardy using a 520mm telescope with a F/D = 50 on Kodak Microfilm exposed for 1 second.*

Photograph 5-13. *Saturn (upper right) photographed by Georges Viscardy using a 520mm telescope with an effective focal length of 24 meters on Kodak 2415 film exposed for 3 seconds.*

Photograph 5-14. *Jupiter (lower left) photographed by Jean Dragesco using a 355mm telescope with a F/D = 50 on Kodak 2415 film exposed for 2 seconds and developed in D19.*

Figure 5-9 gives, for each planet, the relation between the F/D ratio and the exposure time for a film speed of ASA 125, which in white light, corresponds to Kodak 2415 developed in D19.

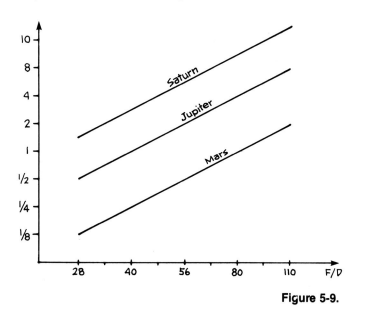

Figure 5-9.

V.4.3. — VENUS

Venus is the example *par excellence* of an object for which filters are indispensable. Of course, you can photograph Venus's phases in white light, but the rare details of its cloudy surface accessible to our telescopes are visible only in ultraviolet.

For Venus, then, the first choice to make is that of a filter. It must let pass ultraviolet and violet light, but block out the other colors. Wratten gelatins, which are economical and usually allow a large range of

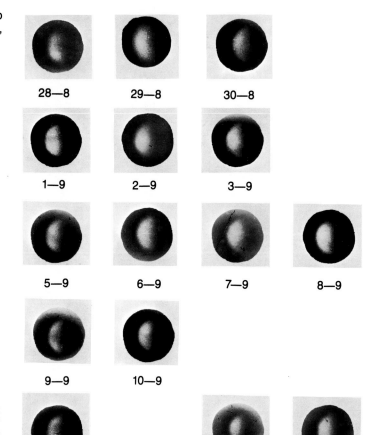

28—8   29—8   30—8

1—9   2—9   3—9

5—9   6—9   7—9   8—9

9—9   10—9

13—9   15—9   16—9

possibilities, unfortunately do not offer a wide choice for these wavelengths. The Wratten 34, 35, and 36 filters can be used, but their transmission factors are a little weak, in spite of a sufficiently large frequency range, and they do not absorb deep red. The Wratten 18A filter is interesting but comes only in glass and costs more. Therefore, in terms of performance, you may prefer instead the Schott's UG 5 or the A358a from MTO.

Moreover, you must beware of the fact that glass—and more particularly the flint element of refractor objectives—absorbs ultraviolet. Flint, which is one of the glasses used in achromatic optics, is opaque to radiation less than 390nm. A reflector is thus preferable to a refractor for photographing Venus. The enlarging lens (Barlow or eyepiece) can also absorb radiation of short wavelengths. Thus, you should take care to use a system in which the glass is the thinnest possible (eyepiece of short focal length, for instance) and which does not contain flint—thus, you will avoid achromatic optics and content yourself with a single-lens Barlow and Huygens eyepieces.

Kodak 2415 is a very good film for Venus, because it is sensitive down to 300nm, in the ultraviolet. You must not develop it at its maximum contrast, however, since the center-edge effect is very noticeable with Venus.

With the UG 5 and the Wratten 34, 35, or 36 filters, it is better to choose an orthochromatic film or a film like Pan F, whose spectral sensitivity is limited to 650nm. In fact, these filters do not absorb red radiation between 650 and 700nm, to which 2415 is still sensitive. The result is the super-imposition of a parasitic red image upon the ultraviolet image.

Since we must photograph Venus in the ultraviolet in order to observe the details of its atmosphere, color films are of no interest.

Photograph 5-17. In 1957, Charles Boyer discovered the four-day rotation of the Venusian atmosphere with this series photographs which show a dark formation that appears one day in four (photos in the first column). He used a 260mm telescope, Kodak Microfilm, a Wratten 34 (ultraviolet) filter and a 1 second exposure.

Because Venus always lies very low on the horizon, the resolution of the photograph is above all limited by atmospheric turbulence. You should thus employ moderate focal lengths and short exposure times.

The absorption of ultraviolet light can vary by very large proportions, due on the one hand to the degree of atmosphere purity (a phenomenon accentuated by Venus's proximity to the horizon) and on the other hand to the optics used. Consequently, it is impossible to recommend precise exposure times and focal ratios. You must thus surrender yourself to a series of preliminary trials.

In addition, the exposures determined by these trials will not necessarily be valuable for all subsequent sessions, because of the great dependence upon atmospheric conditions.

By way of example, the exposure times most often used by Charles Boyer with his 260mm scope varied from ½ sec. to 1/10 sec. (for F/D = 10, a W34 filter and Pan F).

V.4.4. — LUNAR ECLIPSES AND EARTHSHINE

Here are two subjects which often interest the amateur astronomer, although their interest is only aesthetic. Both are instances of photographing the moon when it is not directly lighted by the sun. Naturally its luminous density is low, and the techniques used resemble stellar photography more than planetary photography.

During a penumbral eclipse only a portion of the moon is illuminated by the other portion being hidden by the earth. During a total eclipse the earth masks its satellite from the entire solar disk. A small portion of the sun's light tangential to the terrestrial globe, however, can reach the moon after having been refracted by the earth's atmosphere, which it crosses. Since the short wavelengths are absorbed by the atmosphere, only the red light reaches the moon, bestowing upon it this the familiar reddish hue of total eclipses (Figure 5-10).

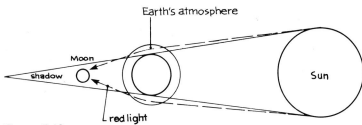

Figure 5-10.

We recommend using a color film to bring out the brown hue of the penumbral phase and the red hue of totality, using a focal length of less than 2.4m (for a 35mm format) to get all of the moon into the frame. Exposure times vary greatly according to the phase. By way of example, we give a few recommended times for an ASA 400 film and F/D = 10:

> full moon: 1/1000 sec.
> mean time for the penumbra: 1/250 sec.
> beginning or end of totality: 4 sec.
> middle of totality: 15 sec.

For different shooting conditions, you need only apply the luminosity equation (see Section IV.7.6) and possibly take into account reciprocity failure (see Section IV.3.1).

At the beginning or end of lunation, the portion of the moon lighted directly by the sun appears as a fine crescent. The other regions, however, are faintly visible to the naked eye—in fact, they return to us a little of the light (called "earthshine") which was previously reflected by the earth. With an ASA 400 film and F/D = 10, you must count on an exposure of 10 to 20 seconds—of course, the crescent will be very overexposed.

In addition to the problems posed by the unreliability of the mounting's right ascension movement during a relatively long shot, you must recall that the moon has a proper motion that can cause a blur if the drive speed is regulated for the apparent displacement of the sky (stellar rate).

CHAPTER VI
SOLAR PHOTOGRAPHY

VI.1. — CHOICE OF PARAMETERS

The sun is the only celestial object for which the problem of too little light does not occur *a priori*. In planetary photography, we must compromise between an effective focal length which is not too large, a sufficiently fast film, and a sufficiently long exposure. These constraints disappear in solar photography: we can let ourselves choose the speed, a very brief exposure, and the focal length we judge necessary.

VI.1.1. — CHOICE OF FILM AND DEVELOPER

In general, solar photography requires a film with a high gamma. In fact, while sunspots are themselves among the most contrasty subjects in astronomy, the same does not hold for the other solar aspects: penumbral details, granulation of the photosphere, faculae.

As luck would have it, the films of highest contrast also have the greatest resolving power. Their single fault—often low sensitivity—is not a handicap here.

Three films are recommended:

1) Kodak's Recordak, developed in D11 or D19. Contrast is excellent and its resolution one of the best you can find. Speed is limited to about ASA 10.

2) Agfaortho 25, ASA 25, which must be used with a paper developer such as Neutol; D19 or Rodinal work equally well. Its peculiarity is that it is orthochromatic, that is, it is sensitive only to radiation of wavelengths less than 600 nanometers (Figure 6-1).

3) The ubiquitous 2415, developed at its maximum contrast in D19.

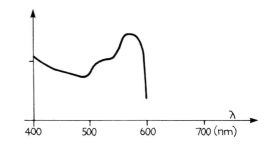

Figure 6-1.

VI.1.2. — EXPOSURE TIME

For conventional cameras the shortest shutter speed is 1/1000th sec., sometimes 1/2000th, or exceptionally 1/4000th (Nikon FE2). The use of these speeds is not uncommon in solar photography. In every case, we must seek to approach these speeds, because the shorter the exposure, the less we need fear turbulence and blurring.

In solar photography the exposure time depends a lot upon the filtration used. If the filtration is severe, you can be led to relatively long exposures (1/30 sec., 1/15 sec., etc.). For speeds of less than 1/60th, you do not have to compensate for the apparent motion of the sky. Thus, a simple altazimuth mounting will suit you; however, you will have to choose one sufficiently robust in order to eliminate vibration problems. For longer exposures the problems found in planetary photography crop up: vibrations, tracking errors, and turbulence. You should know that turbulence is generally much more violent during the day than at night. Consequently, you would be wise to avoid exposures longer than 1/15th or 1/30th sec. The most favorable time for solar observation is usually about two hours after sunrise. The sun is thus high enough above the horizon but has not yet had time to heat the ground sufficiently for turbulence to be important.

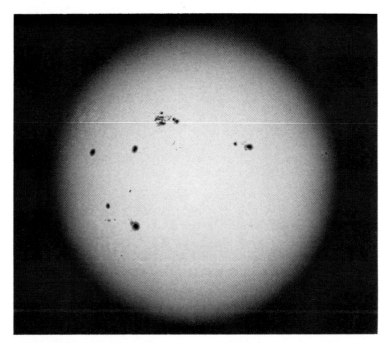

Photograph 6-1. The solar photosphere photographed by F. Rouviere with a 80mm refractor with a effective focal length of 1.2 meters on Recordak AHU 5786 film using an objective filter that transmitted 1/1000 in combination with a green filter (Wratten 54) exposed for 1/125 second and developed in D19b.

VI.1.3. — CHOICE OF FOCAL LENGTH

Just as for the moon, the focal length must be limited to about 2.40m if you wish to obtain a complete image of the sun in 35mm format.

To improve the picture's resolution, you can choose a greater enlargement. The performance of the recommended films, however, lets us use moderate focal ratios. For F/D = 20, the resolving power of the instru-

ment corresponds to 12 μm on the film, which is sufficient. You can, however, employ ratios reaching 30 or 40 so as to make focusing easier and the negative more legible, but in general it is not necessary to go beyond that.

Note in passing that photographing the sun in its entirety on 35mm film (F < 2.4m) while respecting the F/D = 20 criterion is possible with all instruments whose diameter is less than 120mm.

VI.1.4. — ORDER OF PRIORITY

Contrary to what happens with planetary photography, here the exposure time is not of fundamental importance. In fact, it is often less than 1/60th—thus, we can usually ignore the influence of tracking and, in many cases, of turbulence.

So, the most important choice is that of film, and especially of filtration. Next come the enlarging system, and finally the exposure time. You must simply verify that a too restrictive choice of the other parameters does not result in an exposure time greater than 1/30th or 1/15th sec.

VI.2. — FILTRATION PROCEDURES

The sun poses no problem with lack of luminosity, in fact the quantity of light that we do receive from it is so great that filtering becomes absolutely necessary. Depending to the particular case, the proportion of light that the filtration lets pass varies from one hundredth to one ten thousandth of the incident light. Transmitting all the light collected by the objective would have two major consequences: the impossibility of obtaining a sufficiently short exposure time, and the damage due to heating of the equipment where the light is concentrated.

We shall examine the different techniques used for filtering solar light. In no instance should we diaphragm the objective, since that would be translated by a diminution of the instrument's resolving power.

Photograph 6-2. Here (opposite page) the sun is partially shadowed by the large dome of the Nice observatory. F. Rouviere used a 80mm refractor with an effective focal length of 1.2 meters with Recordak film and an objective filter of 1/1000.

VI.2.1. — OBJECTIVE FILTERS

We can use either an opaque filter, which absorbs the light crossing it, or a semi-aluminized glass plate, which reflects most of the incident light. The filtering element is placed at the entrance of the telescope tube, in front of the objective (Figure 6-2).

This solution has considerable advantages:

1) The filter undergoes no important heating. In fact, placed where the light is not yet concentrated, it behaves simply as an object exposed to the sun.

2) The surplus light does not penetrate into the tube of the telescope, where heating in the neighborhood of the focus would inevitably cause harmful turbulence.

3) In the case of an open-tube reflector, the filter also acts as a cover to eliminate a large portion of the tube turbulence.

The drawback to this solution is a financial one. Positioned here, the filter or cover must have the same optical quality as a telescope objective. Moreover, it must have a diameter at least equal to that of the telescope objective. Consequently, a good objective filter can cost more than the mirror it is mounted in front of.

An objective filter is the most efficient solution, however, especially in the case of a reflector. It must have a sufficiently high transmission factor (ideally 1/100; 1/1000 is still acceptable)* uniform over the entire spectrum. Thus we can place farther down the light path a second filter for selecting the desired wavelengths. Conversely, if all the filtration is done at the objective filter (for example, if its transmission factor is equal to 1/10,000) the light transmitted is no longer sufficient to permit a second filtration without falling into the problems of planetary photography. We would therefore have to buy as many objective filters as there are special zones we want to explore.

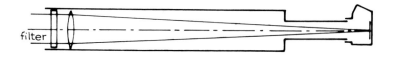

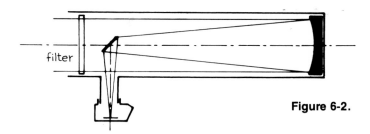

Figure 6-2.

A thin, homogenous sheet of specially coated mylar ($13\,\mu m$ in thickness) can serve as objective filter. This is an economical solution, especially for large diameters, whose quality, however, does not attain that of a glass plate.

VI.2.2. — THE FILTER AT THE FOCUS

Most filters sold for amateur astronomy are designed to be mounted on an eyepiece. This is particularly the case with solar filters furnished with telescopes bought commercially. The natural position for these small-diameter filters is within immediate proximity of the telescope focus, where the concentration of light is the greatest. The heating undergone by a filter at the focus is not only detrimental to image quality because of the turbulence engendered, but is also DANGEROUS: the filter can break and thereby blind the observer.

---

*The sun's angular diameter in radians is 1/100. Thus, at the focus of a telescope with focal ratio F/D and an objective filter of transmission factor T, the illumination equals the direct illumination of the sun multiplied by $T/(10^{-2}F/D)^2$. In order for the illumination to equal the direct illumination from the sun at the focus of a telescope open to F/D = 10, then we must have T = 1/100.

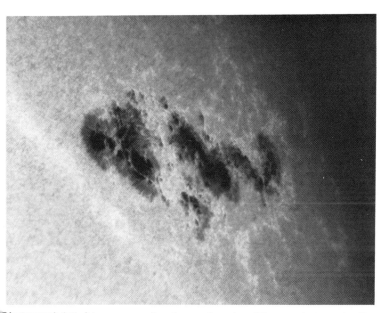

Photograph 6-3. *A large group of spots near the edge of the sun photographed by F. Rouviere. He used a 205mm Newtonian telescope with an effective focal length of 10 meters. An aluminized objective filter reduced the light by a factor of 1/1000. The light was further attenuated by a Schott OG515 filter near the focus. The film was Recordak exposed for 1/250 second and processed in D19.*

The advantage of filters used at the focus is financial. These filters are small and their quality can be mediocre without greatly affecting the image obtained. *You should avoid them, however, if they are not protected by an objective filter or a Herschell wedge, because of the dangers involved.*

Sometimes it is hard to avoid this setup, such as when a photo must be taken in monochromatic light or within a very small frequency range. For this you use a costly interference filter too small to be mounted in front of the objective. Using a neutral objective filter in addition to the monochromatic filter (whose transmission factor is very small) would prove too detrimental to the luminosity. It is fitting, then, to make every effort to avoid excessively heating the filter (limited duration of illumination, positioning the filter on as large a section as possible of the bundle of light, etc.).

VI.2.3. — THE HERSCHEL WEDGE

When you are not using an objective filter, the Herschel wedge is an economical and ingenious solution for limiting the amount of light reaching the filter at the focus. The wedge is placed at a 45° angle in the path of the light rays, a little above the focus. Most of the light passes through the wedge; only a small percentage of it is reflected 90° by the front face. It is this small portion that is used (Figure 6-3). The back face of the wedge is not parallel to the front face in order not to reflect a second image toward the eyepiece tube.

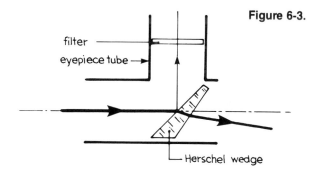

Figure 6-3.

filter

eyepiece tube

Herschel wedge

A wedge is never used alone, since it still reflects too much incident light (coefficient of reflection = 4 to 5%). It is, however, an effective system when used in combination with a filter.

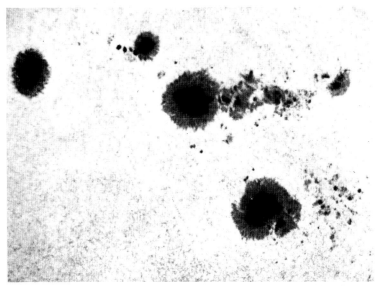

Photograph 6-4. *A group of sunspots taken by Giammertini as part of the observation program sponsored by the Commission du Soleil de la S.A.F. The telescope was a 150mm refractor with an effective focal length of 77 meters using SO115 film exposed for 1/1000 second through a green (Wratten 56) filter.*

VI.2.4. — TAKE CARE OF YOUR EYES!

Solar observation can be dangerous. Too much light hitting the retina can create irreparable lesions and cause blindness. It is imperative for you to respect the following rules of safety:

1) *Use sufficient filtration.* Filters for visual observation of the sun transmit only 1/10,000th of the incident light. With a small instrument (diameter < 100mm) you can let pass 1/1000th of the light collected, but you must not hesitate to attach a supplementary filter behind the focusing system or the focus if you feel bedazzled. Be

aware of the fact that infrared and especially ultraviolet are not visible and thus do not dazzle the eyes, but can cause considerable damage. You must thus redouble your precaution with a filter that attenuates these wavelengths less than those of the visible spectrum. Know the transmission characteristics of your filter!

2) *Never observe behind a filter which can break from overheating.* This is the risk run by filters placed near the focus and not protected by an objective filter or a wedge. The rupture of the filter comes with no warning; the retina instantaneously receives the entire luminous flux.

VI.3. — CHOICE OF FILTRATION

VI.3.1. — REDUCING CHROMATIC ABERRATION

In solar observation refractors are generally preferred to reflectors for the following reasons:

—The sun's strong luminosity does not require a large diameter to collect a maximum of light. Of course the resolving power depends upon the diameter, but it also depends upon the diffraction image and the turbulence.

Now:

—The obstruction caused by the secondary mirror and its support brackets alter a reflector's diffraction image.

—The tube turbulence occurring when you aim at the sun with a reflector that is not closed off with a correction plate or an objective filter can prove disastrous for the images.

—Mirrors are more prone to deformation from heating than refractor objectives.

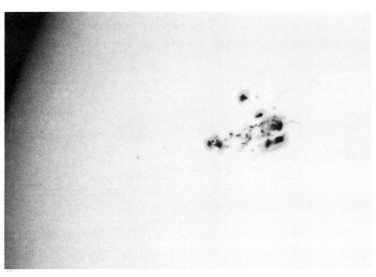

ing solar light gives us a weapon against aberration: when filtering, we must select a band of the spectrum as narrow as possible; thus, the correct focus is unique for all the light arriving on the film.

Apart from interference filters, which are very expensive, selective filters generally have relatively large ranges compared to what we wish to obtain with the sun. Although this can usually be considered sufficient from the point of view of chromatic aberration, we can, however, reduce this frequency range by using two filters, one limiting the window of transmission from the side of the short wavelengths, the other from the side of the long wavelengths (Figure 6-4).

Photograph 6-5. *A group of sunspots taken by the Author using a 90mm refractor, F/D = 26 with an 1/1000 objective filter in combination with a Wratten 29 (red) filter. Exposure 1/1000 second on Kodak 2415 film developed in D19.*

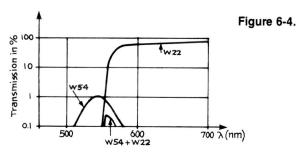

Figure 6-4.

Figure 6-5.

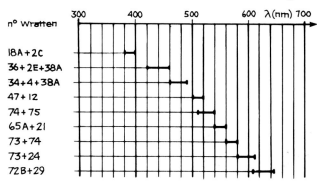

The problem with refractors arises from their residual chromatic aberration, even when the objective is called "achromatic." Recall that chromatic aberration is related to the fact that the index of refraction for glass depends upon wavelength. Consequently, the position of the focus depends upon the color under consideration. Thus, the focusing made for yellow light is not valid for red or blue light. We notice this phenomenon easily with a solar image by making careful focuses with filters of different colors.

When we are obliged to take photos in white light (planetary photography, for example) we can not avoid having the chromatic aberration slightly degrade the image quality. By contrast, the necessity of filter-

Figure 6-5 gives a few examples of the narrow frequency ranges that you can obtain by combining Wratten filters. The Wratten 38A, which occurs in a few combinations in this table, has the purpose of eliminating a second window of transmission in red light. It is not useful if the film used is orthochromatic.

VI.3.2. — THE SEARCH FOR MAXIMUM CONTRAST

Apart from sunspots, details on the solar surface have little contrast in white light. The different zones delimiting their contours, however, have different spectral emissions. When filtering, therefore, you must choose a frequency range which will set into relief what you want to photograph. Here are a few examples.

VI.3.2.1. — VIOLET, BLUE, AND GREEN LIGHT

The theory of black bodies (which the sun roughly obeys) teaches us that the material emits light according to its temperature. The hottest body emits the most light and the maximum of its emission lies in the shortest wavelengths (Figure 6-6). Thus it seems that the maximum luminous contrast between two emitting bodies of different temperatures lies in the short wavelengths.

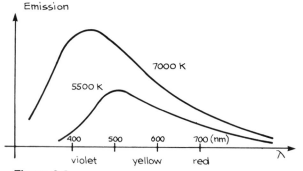

Figure 6-6.

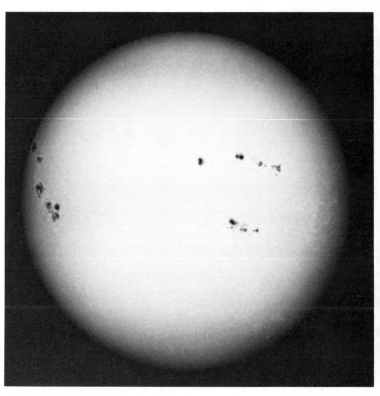

Photograph 6-6. Sunspots and faculae on the sun's surface by Francis Grase using a 81mm refractor with an effective focal length of 2.2 meters, and an 1/1000 objective filter on Agfaortho 25 film for 1/500 second.

Photograph 6-7. The photograph on the next page, upper left, is of a group of sunspots near the sun's edge was photographed by the Author using a 90mm refractor with a F/D = 26, a 1/1000 objective filter in combination with a violet (Wratten 34) filter on Kodak 2415 film exposed for 1/250 second and developed in D19.

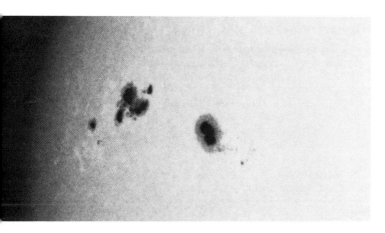

In the sun's case, the faculae—areas which are hotter and therefore more brilliant than the rest of the photosphere—are more apparent in violet, blue, or green light than in yellow or red light. Atmospheric diffusion, however, which becomes greater the shorter the wavelength, tends to eliminate this gain in contrast. A good compromise, then, consists of photographing the sun in green light.

We can recommend the following filters. The figures in parentheses indicate the wavelength of the maximum of the window of transmission:

Violet light: Wratten 34 (420nm), 35 (420nm), 36 (410nm).
Blue light: Wratten 47 (440nm), 44 (490nm).
Green light: Wratten 40 (510nm), 55 (520nm).

A word of warning: most of these filters have a second frequency range for wavelengths longer than 600 or 650 nanometers (red light). A red image can thus superimpose itself upon the desired blue or violet image, which is especially annoying from the point of view of chromatic aberration, since focusing for these two colors can differ. To avoid this inconvenience, you should use a film insensitive to these long wavelengths, such as Agfaortho 25, or add a second color filter opaque to red light.

Unfortunately, a filter of satisfactory transmission for the short wavelengths and perfectly opaque to red does not exist. Instead, you can use the Wratten 38 or 38A.

If you are content with a relatively large spectral band—from violet to yellow—you can use Agfaortho 25, thanks to a chromatic sensitivity restricted to short wavelengths ($\lambda < 600nm$) in the absence of any color filter. Conversely, a narrow frequency range can be obtained by combining two filters.

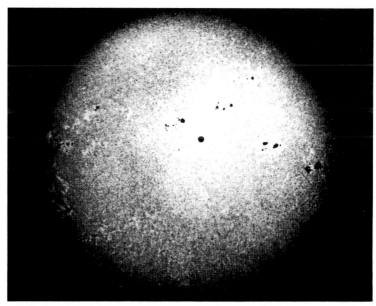

Photograph 6-8. *This photograph of the solar chromosphere for the calcium lines H and K was taken by Giammertini using a 200mm Schmidt-Cassegrain telescope with a F/D = 13, a Wratten 18A filter and a 1/250 second exposure.*

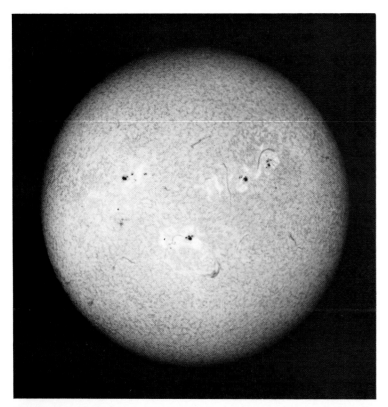

Photograph 6-9. *Jean-Marie Roques took this photograph of the solar chromosphere in Hydrogen alpha light using a 75mm diameter refractor with an effective focal length of 2,177mm on SO392 film and a filter with a 1 angstrom frequency range. Note the filaments — protuberances seen on the disk — as well a faculae and chromospheric granulation.*

VI.3.2.2. — CALCIUM LINES H AND K

The calcium lines H and K, situated at 393 and 396nm, at the limit of ultraviolet and violet, are important absorption lines of the solar spectrum. The photosphere thus hardly appears in these wavelengths. On the other hand, photography in the calcium lines lets you visualize the inner chromosphere. The faculae are thus very high in contrast and visible even at the center of the solar disk.

For want of an interference filter—the ideal but costly solution—you can obtain very good results with a Wratten 18A (green filter) and a Wratten 2B which delimit a frequency range from 390 to 400 nanometers. If the film used is Kodak 2415, you will have to add a Wratten 38A, which eliminates the second window of the W18A lying beyond 700nm. The effectiveness of this method has been demonstrated recently by an amateur astronomer from Nice, Thierry Laverge.

VI.3.2.3. — THE Hα LINE OF HYDROGEN

This is another absorption line in the solar spectrum. The Hα line, situated in the red, gives an image of the chromosphere at about 2500km in altitude.

Contrary to the calcium lines, the Hα line, formed in the red around $\lambda = 6563$ Å, is very narrow (only a few Å). It is thus necessary to use an interference filter with a very narrow frequency range (4 to 5 Å maximum).

VI.4. — SOLAR ECLIPSES

VI.4.1. — PARTIAL AND ANNULAR ECLIPSES

To the extent that a portion of the photosphere remains visible, we can not observe the solar atmosphere. The only possible photographs consist of fixing the contour of the moon standing out in silhouette against the luminous surface of the sun.

In fact, it is quite simply a question of photographing the photosphere and the conditions of photography are exactly those that we have described previously.

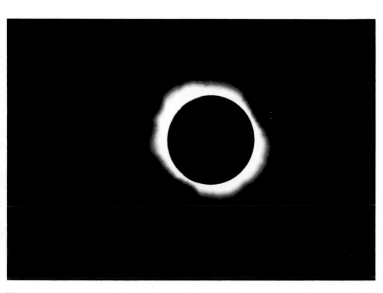

Photograph 6-10. *Total eclipse of the sun on 7/31/81, photographed by Honore Arioli at Polyssajevo (Western Siberia) with a 800mm telephoto, a F/D = 8 on Ekatachrome 200 exposed for 1/500 second.*

VI.4.2. — TOTAL ECLIPSES

This phenomenon is much more interesting, but alas, more rare. The total disappearance behind the moon of the source of intense light constituting the photosphere allows us to catch sight of the solar corona—hence, the interest in total eclipses.

VI.4.2.1. — FOCUSING CONDITIONS

Given that we are interested in the neighborhood of the solar disk, the field of interest is relatively extended. Consequently, we can not use too long a focal length. The ideal equipment is a large telephoto or a small telescope, whose focal length lies between 400mm and 1200mm. Of course, the objective must be used at full aperture.

The film to use is Kodachrome 64 if the objective is open between $F/D = 5$ and $F/D = 8$, and Ektachrome 200 if the ratio lies between 10 and 15.

The solar corona presents a large variation in luminosity. The very luminous inner corona necessitates exposure times between 1/1000th and 1/250th sec. By contrast, the outer corona requires times ranging from 1/60th to 1 sec. Since total eclipses are very rare phenomena, you must make a wide range of exposures to explore the entire breadth of the inner and outer corona.

You should also take care to center the sun well within the photographic field, especially for images of the extended corona. A motorized equatorial mounting can be useful, especially for long exposures. The desired resolution for the corona, however, can tolerate the slight displacement owing to a fixed mount. Besides, an equatorial mounting is a heavy piece of equipment that can be difficult to transport to the eclipse site. Don't forget that you should be able to adjust your mount for the latitude of the site.

If the support for the focusing apparatus is scanty (simple photographing support, for example) you should be very vigilant about the problem of vibrations. If possible, use a camera whose mirror can be raised manually and wait a few seconds between each shot.

VI.4.2.2. — PLANNING THE PHOTOS

The phase of totality for a solar eclipse is very short. The exact time varies and can be found in ephemenidies. You must therefore avoid any loss of time and prepare your observation plan in advance, according to the equipment at your disposal. A motor drive for your camera can be useful.

The first picture of this phase must be taken at the exact moment of the second contact, at the disappearance of the last ray of sunlight. This is very important, for it is the sole instant and the sole location (where the contact is made) where the lunar disk hides the solar disk exactly without jutting out. After second contact, you can no longer perceive the inner corona at the solar limb level, the lunar disk being larger than the solar disk. The phenomenon recurs at the precise moment of the third contact, just

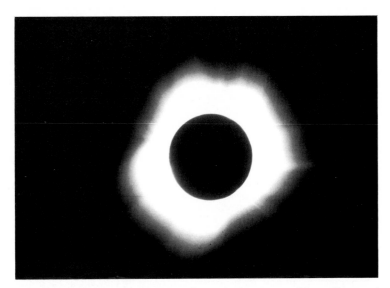

Photograph 6-11. *This photograph was taken under the same condition as Photograph 6-9 but the exposure was lengthened to 1/15 second.*

before the first ray of sunlight reappears, but this instant is harder to pin down. This first photograph must be taken at 1/1000th sec. followed immediately by another.

The rest of the program must follow it without loss of time. We advise you to explore progressively every exposure time, going from 1/1000th to 1/250th sec., then from 1/60th to 1 sec.

It can be useful to take photos of the solar disk before and after totality. These allow you to mark the position of sunspot groups. The groups near the limb may exist in correlation with the protuberances perceived during totality.

VI.5. — CORONOGRAPHS

The rarity of total eclipses of the sun has driven astronomers to create "artificial eclipses" in order to observe the corona regularly with the help of special instruments called *coronographs*. The first coronograph was installed at Pic du Midi in 1930 by the French astronomer Bernard Lyot.

The principle behind the coronograph is simple. In Figure 6-7, objective Ob forms an image of the sun at its focus F, where it is occulted by the disk D. This disk carries a cone which redirects the light along the sides. Since the diameter of the disk is slightly greater than that of the sun's image, the corona can be observed on all sides of it. The collimating lens C, whose focus also lies at F, gives an image at infinity of the occulting disk and of the corona. This image crosses a diaphragm d whose role is to stop the solar light diffused by the edges of the objective Ob, and an interference filter f. The image is then recaptured by the objective Oc of the camera, which focuses it on the film.

Figure 6-7.

The problem with the coronograph stems from the fact that the solar disk is a million times more luminous than the corona. The light diffused by the edge of the objective is no longer negligible, hence the existence of the diaphragm d. The objective itself must be ground to perfection from a glass of great homogenity. But above all, the purest terrestrial

Photograph 6-12 a, b, c. *This series of photographs was taken by A. Doucet with a 110mm coronograph he built himself. Interference filter with a 4 angstrom range centered on 6563 angstrom.*

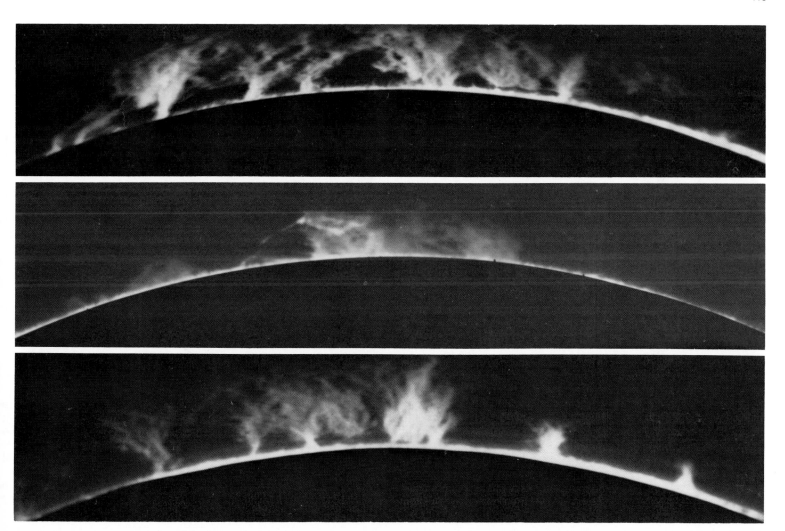

atmosphere diffuses around the sun ten times more light than the corona itself, which is thus found masked—hence the necessity of using the coronograph on a very clear site (a mountain) and of installing an interference filter which lets pass only the emission wavelengths of the corona, thus filtering out the atmospheric "noise."

In spite of all these precautions, it is impossible to observe the solar corona as well as during a total eclipse. Besides, the construction and utilization of a coronograph are so difficult that very few amateurs have taken on the challenge.

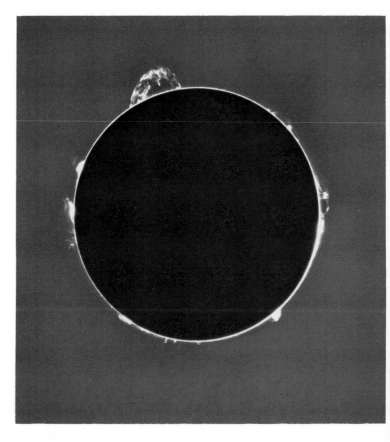

Photograph 6-13. Dany Cardoen took this photograph of the solar chronosphere with a coronagraph of F = 1,500mm which he built. The exposure was made on TP2415 at 1/8 second.

CHAPTER VII

FAINT-OBJECT PHOTOGRAPHY

VII.1. — GENERALITIES ON FAINT OBJECTS

Under the classification "faint objects" we group all celestial bodies requiring relatively long exposures, from one minute to a few hours. This category, then, is a particularly rich one, since we find in it everything that can interest the amateur astronomer, apart from the sun, the moon, and the principal planets. Objects qualified as faint objects include stars and star clusters, galaxies, nebulae, comets, etc.

The luminosity scale for these objects is very extended, since from the most brilliant stars to the faintest nebulae scarcely detectable in the depth of the sky there exists an intensity ratio of 10 billion. Yet in every case, the length of the exposure time forces the operator to concentrate on criteria differing from those of planetary photography.

Moreover, you can not employ these long exposures without penalty. One consequence is that in most cases you will be unable to attain the theoretical resolution of the optics used, contrary to what happens in planetary photography.

The characteristics of films lead us to make two distinctions among faint objects. The first concerns the apparent extension of the object in the sky. Nebulae and galaxies having measurable apparent dimensions are qualified as extended objects, as opposed to stars, which are perceived as point sources, whatever the telescope used.

The second distinction has to do with the contrast in luminosity between the object of interest and its environment. We can form two classes of detection along these lines. Class I contains the objects that are clearly visible with respect to their surroundings, even if they are faint. This includes stars of relatively low magnitude, of nebulae much more luminous than the

Photograph 7-1. The galaxy M31 in Andromeda and its companion M32 taken by Jacques Silvain with a 158mm telescope, F/D = 4.5 on 103aO film exposed for 30 minutes.

background sky, etc. By contrast, Class II consists of those objects scarcely emerging from the ambient light, volutes of gas in a nebula, etc.

VII.2. — EXPOSURE TIME

VII.2.1. — EXPOSURE AND RESOLUTION

We saw in planetary photography that the exposure should last less than a second, in order to avoid the perturbation caused by atmospheric turbulence. In faint-object photography, it is out of the question to respect such a constraint. The picture will thus be limited in resolution by the amount of turbulence existing at the moment the photo is taken. This limit can vary from 1″ to 10″, according to the site and the condition of the atmosphere. The influence of turbulence, however, is the same, whether

Photograph 7-2. *The Lagoon nebula, M8, by Rene Gouzy with a 200mm telescope, F/D = 5 on 103aF film exposed for 45 minutes.*

the exposure takes 30 seconds or 2 hours. Turbulence in faint-object photography must therefore be considered as a limitation of the resolution and not as a limitation of the exposure time, contrary to what happens in planetary photography.

The same holds true for the vibrations to which the telescope can be subjected. If their amplitude is greater than that of the turbulence, they will have to be eliminated by the reinforcement of the mounting. In fact, it is no longer possible to place the exposure within the brief interval separating two gusts of wind, for example.

The third relation between exposure time and resolution comes from the tracking of the sky's apparent motion done by the equatorial mounting. In planetary photography this criterion also resulted in a limitation of the exposure time, on the order of a few seconds for a well-adjusted mounting. It is thus out of the question to take an exposure lasting several minutes without controlling the tracking. Even if the mounting were perfect, it would produce a slight displacement of the body photographed with regard to the direction sighted by the telescope. In fact, in the course of a long exposure the height of celestial bodies from the horizon varies appreciably, and with it the value of atmospheric refraction. The operator must therefore guide the telescope during the entire exposure.

VII.2.2. — GUIDING

Guiding the exposure presupposes on the one hand possible measures you can take with the mounting, and on the other hand a way of controlling the effectiveness of these measures.

The mounting must be equatorial and equipped with a motor—it is not possible to do proper tracking manually, even at the focus. But this is not enough: you must be able to accelerate or decelerate the rotation about the polar axis, in order to compensate for the irregularities in the drive mechanism. You can either change the speed of rotation directly with an oscillator drive if the motor is synchronous, or superimpose a second movement with a differential gear, for example. Likewise, you must be able to make small "corrections" in declination in order to compensate for an inexact alignment of the polar axis. These adjustments, in right ascension as well as in declination, must be sufficiently gentle to be compatible with the precision sought.

Controlling the telescope's displacement with respect to the sky is done with the help of a reticulated eyepiece. Such an eyepiece has a crosshair or reticle engraved on the glass in its front focal plane. Thus, this reference reticle seems sharp at the same time as the stars in the field. In order to be visible against a dark sky, the reticle must be illuminated by a side light. This light can be varied in order to adapt its intensity to the luminosity of the star you have chosen for a guide star (Figure 7-1). Using these aids, the operator must keep the guide star in a fixed position inside the reticle lines. The scope then remains directed toward the same point in the sky.

Figure 7-1.

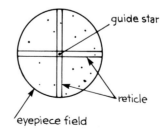
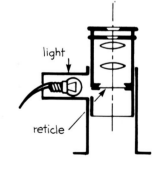

Guiding precision—and thus the perception of the star's displacement with respect to the reticle—must be such that the maximum blur is less than 30 microns, the resolving power of fast films. Now, since the lines of a reticle are usually spaced 0.1mm apart, 30 microns represents just the resolution that the eye can attain at the focus of a reticulated eyepiece. So that the angular displacement perceived by the eye may be compatible with the blur tolerance, the focal length of the objective serving as guide should be at least equal to the focal length of the photographic objective.

Three possible solutions exist:

1) Surveillance is done by a second telescope of a focal length close to the first, placed parallel and having a reticulated eyepiece. This is the most practical solution, but it requires two telescopes and a sturdy mounting in order to carry them simultaneously (Figure 7-2).

2) A part of the light coming from the photographic objective is directed toward the guiding eyepiece by a semi-reflecting plate (Figure 7-3). The plates commonly used redirect 10 to 20% of the incident light toward the eyepiece. Although this amount of light proves insufficient for tracking when the guide star is faint, there is the additional inconvenience of transmitting toward the film only 80 or 90% of the light gathered by the objective.

3) A mirror at 45° or a prism placed a little in front of the camera, on the edge of the field, redirects the image of stars lying on the edge of the zone photographed toward a guiding eyepiece on the side (Figure 7-4). Compared to the preceding solution, this one has the advantage of furnishing the eyepiece with a luminous image without decreasing the illumination of the film. On the other hand, the zone in which you must choose the guide star is limited and you can have trouble finding one bright enough.

The recent use of Kodak 2415 imposes a greater restriction upon the guiding. This film can resolve details of $10\,\mu$m, three times finer than other films. The preceding reasoning leads us to using a guide scope whose focal length is triple that of the photographic telescope, which is often difficult. One good solution, then, consists of keeping one of the three setups just described and attaching a Barlow lens in front of the reticulated eyepiece.

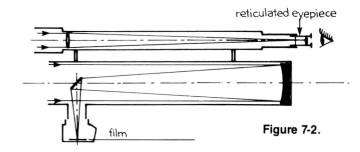

reticulated eyepiece

film

Figure 7-2.

Figure 7-3.

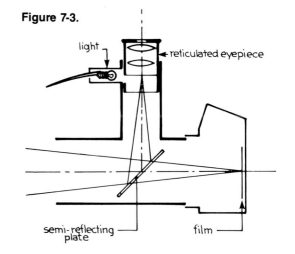

light

reticulated eyepiece

semi-reflecting plate

film

Figure 7-4.

light → eyepiece

→ reticle

prism

film

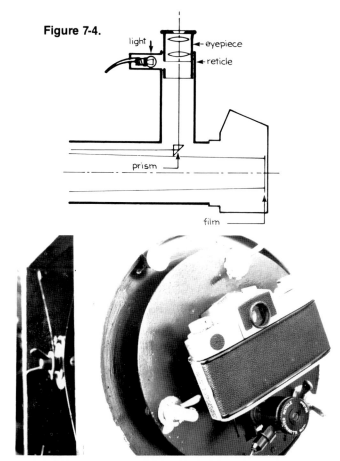

Comets are faint objects with the peculiarity of having a generally significant proper motion with respect to the stars. The guiding, then, is done with the comet itself and not with a guide star. Not only does this exclude the guiding method using the edge of the field—since the image of the comet can not be directed to both the film and to a guiding eyepiece at the same time—but above all, tracking is difficult because it is hard to assess with precision the comet's position with respect to the reticle, the comet being an extended, fuzzy object. Another solution consists of choosing a guide star in the neighborhood of the comet and—the eyepiece being equipped with a micrometer—displacing the star in the eyepiece field while the exposure takes place, in order to compensate for the comet's motion. Of course, this method assumes that you have calculated the comet's speed of displacement in advance from its ephemerides . . . and that you do not make a mistake in direction while making your corrections!

Photograph 7-3. Gilles Calvet made this combination adapter-guider. The illuminated reticle eyepiece is below the camera and receives star images from the edge of the telescope's field of view by means of two reflecting prisms. Note the three turn-screws that allow "squaring-on" of the cameras' film plane to the telescope's focal plane.

Photograph 7-4. Comet Austin taken 8/21/82 by Christian Viladrich using a 200mm Schmidt-Cassegrain telescope with a focal reducer, F/D = 5 on 103aO film exposed for 5 minutes.

VII.2.3. — MAXIMUM EXPOSURE

We have just seen that the limitations to exposure that we studied in planetary photography do not apply here. Of course, the exposure will have to be adapted to the film speed, to the focal ratio, and to the luminosity of the object photographed, so that the film will be correctly exposed, that is, so that it has the desired density.

But for the same film density, a long exposure allows a larger focal ratio and hence a longer focal length. Now, a long focal length improves resolution, since it provides the film with a larger image. Consequently, we shall always choose the longest exposure possible.

Finally, the exposure time in faint-object photography is limited only by the skill—and especially the endurance—of the operator. In fact, guiding is a difficult operation, demanding training and a lot of attention. When the guide star is faint, eye fatigue becomes considerable. Most stellar photos do not exceed an hour in exposure, since guiding the scope for much longer turns into a sporting event.

VII.3. — RECOMMENDED FILMS

We know that film speed rapidly decreases for long exposures—reciprocity failure sets in. For example, with Tri X, which for exposures of less than a second has a speed of ASA 400, the speed plummets to about ASA 5 at the end of a half-hour exposure. Thus, you need to use either films corrected for reciprocity failure or films hypersensitized beforehand.

VII.3.1. — THE 103a AND IIIaJ TYPE FILMS

We saw in Section IV.4.6 that these are a series of Kodak films corrected for reciprocity failure and available to amateurs. They enjoy the double advantage of having a rather high contrast and coming in different spectral sensitivities. The disadvantage of the 103a films is a relatively high granularity, which considerably limits the resolution of photos with short focal lengths. IIIaJ's low sensitivity necessitates long exposures and short focal ratios. Although it is already corrected for reciprocity failure, you would do well to hypersensitize it.

VII.3.2. — HYPERSENSITIZED FILMS

A correctly hypersensitized film keeps its nominal speed without a degradation in performance during a long exposure. The various techniques have been explained in Section IV.3.3. All films can undergo this treatment with varying results; thus, here we shall consider the one giving the best performance. As we have seen previously, the prize must go to Kodak 2415.

Hypersensitized Kodak 2415 is slightly more sensitive and much higher in resolving power than the 103a films. Moreover, 2415 has a better signal-to-noise ratio than the 103a films, which makes it more powerful for detecting low-contrast objects. Thus, it lets you gain about 2 magnitudes in detecting limit stars against the background sky. Kodak 2415, then, is clearly preferable to 103a. The only drawback lies in the fact that hypersensitization is not a flexible technique: it requires special equipment, usually must be done just before shooting and processing should follow immediately after exposure.

VII.3.3. — COLOR FILMS

These are not designed for long exposures. The decrease in speed and contrast owing to the Schwarzschild effect does not occur with the same intensity in the three sensitive layers composing the films. Each of these layers represents a primary color (blue, green, red). Consequently, the colors obtained on a photo exposed for a long time are inexact.

To limit these problems, you can hypersensitize color film in Forming Gas, or expose it with a cold camera.

The reversal film Fujichrome 400 enjoys the advantage of a moderate granularity and a high speed. Its performance improves with hypersensitization in Forming Gas, or with negative development. You should choose it for short focal length photos, especially for objects emitting lines in the red ($H\alpha$).

Photograph 7-5. The globular cluster M22 photographed by Serge Deconihout with a 200mm Newtonian on 103aF film exposed for 1 hour.

Photograph 7-7. *The Dumbbell nebulae, M27, by Christian Arsidi using a 280 Schmidt-Cassegrain telescope with focal reducer, F/D = 5 on Kodak 2415 film hypersensitized in forming gas exposed for 1 hour.*

Photograph 7-6. *The Veil nebula, NGC 6990, by Christian Viladrich with a 200mm telescope, F/D = 5 on 103aF film exposed for 1 hour. Note that in this and Photograph 7-5 the stellar images are perfectly circular, indicating perfect guiding despite a long exposure.*

Photograph 7-8. *Christian Arsidi took this picture of the galaxy M51 using a 280mm telescope, F/D = 5, on 103aO film, exposed for 40 minutes. The relatively large grain of 03a is clearly visible here.*

The 3M ASA 1000 film is useful for pictures when a short exposure is desired, without hypersensitization, or when you want to attain a high stellar magnitude without worrying too much about a large granulation. As for Kodacolor VR 1000 negative film, it is better adapted to faint extended objects than to stellar photos.

VII.4. — FILTERS

In faint-object photography filters can play very important and varied roles. You must be aware, however, that a filter always lowers the film's illumination and thus leads to an increased exposure time.

VII.4.1. — U B V PHOTOMETRY

This operation consists of photographing stellar fields in three spectral zones: band U ("ultraviolet") obtained with 103aO and a UG 11 filter; band B ("blue") obtained with 103aO and a GG 385; and band V ("visible") more exactly centered in the yellow, obtained with 103aD and a GG 495. A photometrist perusing these three pictures will know the luminosity of the stars in each of the bands. The comparison for each star among these three measurements lets him know its surface temperature.

Notice that a photo in blue light brings out the peripheral regions of galaxies, composed of hot yellow stars, while a photo in red light (103aE, 103aF or 2415 with a W25 or W29) favors the core, constituted of older, cooler stars.

VII.4.2. — THE THREE-COLOR PROCESS

Here we also take three pictures of the same object with three different frequency ranges, but in this case our aim is to obtain a color composite. The three regions of the spectrum chosen are the three primary colors: blue, yellow, and red. Two filtering systems are used, one employing 103a, the other Kodak 2415:

blue: 103aO + GG 385	or: 2415 + W47B
yellow: 103aD + GG 495	2415 + W58
red: 103aF + RG 610 or W29	2415 + W25.

Photograph 7-9. The nebula M17 by Christian Viladrich with a 200mm telescope, F/D = 5 on 103aF film, orange (Wratten 12) filter exposed for 30 minutes.

Photograph 7-10. The Pleiades (opposite page), an open cluster taken by Jean-Marie Roques with a Schmidt camera, 210/310/490, on IIaO film exposed for 11 minutes. IIaO's spectral sensitivity, limited to the short wavelengths, permits a record of the clouds surrounding these stars without having to use a blue or violet filter.

Photograph 7-11 and 12. These two pictures (overleaf) were taken by Jean-Marie Roques with a Schmidt camera, 210/310/495, on 103aF film. At left, the California nebula, NGC 1499, in Perseus was exposed for 1 hour 16 minutes through a red (Wratten 29) filter. At right, nebulosities near nu Cyg and NGC 6888 was exposed for 46 minutes through a red (Wratten 12) filter. The use of these filters allows us to highlight clouds rich in hydrogen which radiate a great deal in red.

Thanks to the spectral selectivity of the 103a type films, the first system uses filters that are not very severe; thus, the film's illumination is not too diminished. On the other hand, the W47B and the W58 necessary with 2415 are very selective. The second system, therefore, can not be used realistically except with optics having very open f-ratios—Schmidt telescopes or small telephotos. The interest in this system resides in the fact that 2415 furnishes images of much higher resolution than the 103a films.

When you have obtained the three black-and-white negatives of the same object, you need only print them one at a time on color paper, using the same filter through which you exposed the negative. The color image is thus recomposed.

The advantage of the three-color process is the use of more powerful films than color films. Besides, there is no longer the problem of chromatic balance, since the operator can adjust the exposure times, keeping in mind the spectral sensitivity of the films and of the transmission factor for each filter.

VII.4.3. — SELECTING EMISSION LINES

Nebulae are gaseous bodies emitting light of quite precise wavelengths: these are the emission lines of the gases that constitute them. Selecting one of these lines with a filter lets you determine the extension of the gas concerned, or more simply, to better bring out a nebula whose contrast to the background sky is low (the Horsehead Nebula or the Veil Nebula, for example). The loss of light from the filter is not considerable, since most of the radiation lies within a few principal lines.

Each line can be selected with a interference filter of narrow frequency range, but this is expensive. The Hα hydrogen line, however, situated at 656 nanometers in the red—which is one of the most remarkable emission lines for nebulae—can be sufficiently isolated simply with 2415 and a W29 gelatin filter.

VII.4.4. — FIGHTING LIGHT POLLUTION

Increasingly intrusive urban lighting tends to make the night sky brighter and brighter and consequently the faint nebulae lower and lower in contrast.

Selecting a spectral emission line of the nebula lets you recover (at the expense of an increased exposure time) a contrast compatible with the film's detection. Special filters called LPR's (Light Pollution Rejection) come with two frequency ranges: one in the red where Hα lies, the other around 500 nanometers, where we find three other important lines: Hβ and the two "forbidden" lines of doubly-ionized oxygen (O^{III}). On the other hand, these filters are opaque to the other colors in the spectrum, particularly to the emission lines of sodium. Since many urban areas use sodium lighting it is a major component of the background sky and can be reduced with the proper filter.

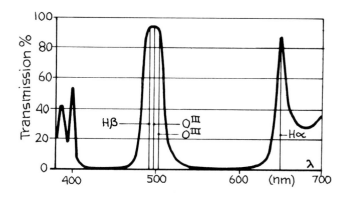

Photograph 7-13. The Trifid nebula (M20) by Jean-Paul Trachier with a 200mm telescope, F/D = 5, on 103aF film exposed for 35 minutes.

VII.5. — FOCAL RATIO

We have just chosen the films to use (103a or 2415 hyper-sensitized), possibly with a filter, as well as the exposure (the maximum the operator can tolerate). It remains for us to define the proper focal ratio for obtaining the desired film density.

VII.5.1. — FOR EXTENDED OBJECTS

The image surface of such an object on the film varies as the square of the focal length and hence of the ratio F/D; the illumination of the film is inversely proportional to the image surface and hence to the square of the ratio F/D. Generally, you will balance the longest focal length (for maximum image size) against the longest possible exposure (for maximum resolution). In fact, you can not exceed F/D = 10 to 15 even for the most luminous bodies.

Note that with Kodak 2415, it is theoretically useless to exceed F/D = 17, the value for which the telescope's resolving power corresponds to an image of $10\,\mu$m, comparable to the resolving power of the film. Likewise, you should take care not to exceed a focal length of 2m, for which an image of $10\,\mu$m represents an angular diameter of 1″, which is the limit of resolution imposed by turbulence.

If the object photographed belongs to Class II—that is, if it is embedded in an ambient luminosity comparable to its own—you have to expose for the best signal-to-noise ratio of the film. We know that this optimum is attained for a negative density close to 1 (compare the negative visually with a neutral filter of density 1 to insure that the exposure is correct). Now, such a density can be obtained in a reasonable exposure time only if the focal ratio is low enough (less than 5).

VII.5.2. — FOR STARS

A star is a point object; consequently, the image formed from it on the film corresponds to the telescope's diffraction disk. Now, for focal ratios less than 8, this diffraction image is less than $5\,\mu$m. Because of the diffusion of light in the gelatin, the stellar image on the film can not sink below this value. A focal ratio less than 8, therefore, does not augment the illumination of the film for a stellar image. On the other hand, it does augment the illumination for the background sky fog which must itself be considered as an extended object. For the detection of faint stars, it is thus desirable to use a focal ratio between 5 (optimum signal-to-noise ratio of the film) and 8 (optimum ratio of star-to-fog illumination).

VII.5.3. — ADAPTED TELESCOPES

In the actual practice of faint-object photography, the margin of maneuverability for choosing the focal ratio of a given instrument is very limited. Most refractors have primary focal ratios greater than 10, which limits their use. The commercial Schmidt-Cassegrain can be used at the

Photograph 7-14. *Spiral galaxy, NGC 2903, in Leo by Herve Le Tallec with a 200mm Schmidt-Cassegrain telescope, F/D = 10 on 103aO film exposed for 1 hour.*

Photograph 7-15. *The globular cluster, M13, in Hercules by members of the Amateur Observatory of Triel with a 310mm Newtonian telescope, F/D = 7 on 103aO film exposed for 45 minutes.*

Cassegrain focus with F/D = 10, or F/D = 5 with a focal reducer. As for Newtonian telescopes—whose primary focal ratios are generally between 5 and 8—they could also be used at between 10 and 16 with the aid of a Barlow. A Newtonian telescope whose focal ratio is less than 5 presents a strong coma as soon as you move from the center of the field. In order to remedy this, you would have to mount a field corrector above the focus. This is an accessory that you must make yourself, since it is not found commercially.

The best solution for attaining focal ratios less than 5 is the Schmidt camera, which has an extended coma free field. Schmidt cameras can be used only at the primary focus and at their nominal focal ratio.

For short focal lengths (less than 500mm) you can procure sufficiently fast telephoto lenses (f2.8 to f4.5) which can be highly useful in faint-object photography. They are easier to use, but much more expensive than Schmidt cameras with comparable characteristics.

As we have seen through this short theoretical review, it is often impossible to apply the desired focal ratio to the particular equipment you have. This theory, however, helps direct astronomers toward the purchase of the type of instrument which is best adapted to the photography they want to do. This instrument is generally used at its nominal focal ratio (prime focus photography). We advise you, then, to adjust the exposure time according to this ratio.

Moreover, a certain number of faint objects extend over several degrees: fields of galaxies, large nebulae, comets, etc. You must then choose an instrument with the proper field: telephoto, Schmidt camera, or "Flat Field Camera" (the FFC is a sort of telephoto whose optics are close to that of the Schmidt-Cassegrain).

CHAPTER VIII
SOME TECHNIQUES TO KNOW

VIII.1. — ALIGNMENT

We have noted the importance of aligning the telescope's polar axis with the earth's axis. We shall now describe several methods for making this alignment.

VIII.1.1. — SIGHTING THE POLESTAR

Fortunately there is a bright star close to the celestial pole. This is the famous "North Star," α of Ursa Minor (α UMi).

We can therefore align the scope by sighting α UMi with a guide on the mounting that will establish its polar axis. In order to facilitate this operation, some telescope manufacturers have incorporated a small sighting scope, within the mounting's polar axis.

You must remember, however, that α UMi does not lie exactly at the celestial pole, but about a degree from it. Figure 8-1 indicates the pole's location with respect to neighboring stars so that you can make the necessary correction following the sighting of α UMi. Sighting Polaris is the simplest and fastest method, but it is not very precise when done like this.

Figure 8-1.

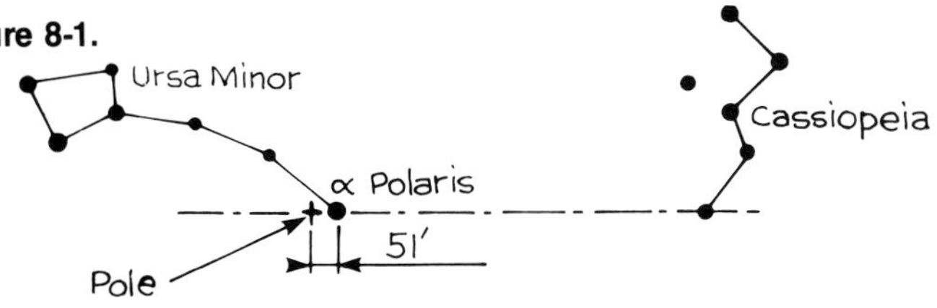

One good solution for improving precision in sighting Polaris consists of using the telescope itself for finding the pole. This method works in two steps. First, you align the telescope's optical axis with the instrument's polar axis, setting a declination of 90°. Second, the right ascension and declination movements being locked, bring the celestial pole to the center of the telescope field (which aligns the instrument's polar axis) while moving the entire mounting.

This method differs little from simple polar sighting. The sighting is simply much more precise because it is no longer done with a mechanical guide, but with an optical instrument magnifying some tens of times.

On the other hand, the visible field is narrower (being that of the telescope) and you must have a good map or know the region of the celestial pole very well.

The first step, which consists of aligning the scope's optical axis with the mounting's polar axis, can be done this way: after an approximate aiming of the tube, turn the instrument about its polar axis. At the eyepiece, you will observe that the stars in the field describe arcs of circles whose center is none other than the direction sighted by the mounting's polar axis. All you need do, then, is play with the declination movement to bring this point into the center of the field.

With a little care and practice, this polar sighting of the scope can be done in about ten minutes, allowing a precision of 5 to 10 minutes of arc. You can't do this, however, except with mountings that let the telescope tube aim in the direction of the pole.

VIII.1.2. — BIGOURDAN'S METHOD

This method relies upon the fact that a bad aiming of the polar axis causes a drift in declination of the body being followed.

Imagine that the instrument's polar axis lies well within the scope's meridian plane, but is too inclined with respect to the zenith (Figure 8-2).

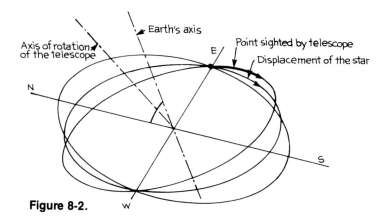

Figure 8-2.

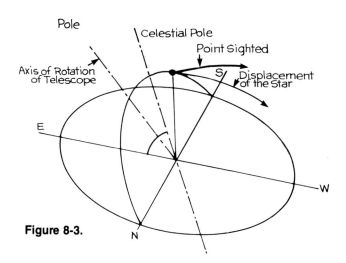

Figure 8-3.

Let's aim at a star in the East and follow its displacement. Since the telescope's axis of rotation is more inclined than the earth's axis, the point sighted follows a path that "ascends" more rapidly than the star. The star, then, moves toward the base of the field of view. The observer thus sees it rise with respect to the eyepiece reticle, since telescopic vision is inverted.

The first rule of the Bigourdan method is: if a star in the East "ascends" the eyepiece field, it does so because the polar axis is too inclined. The opposite conclusion applies, of course, if the star "descends" or if it lies in the West.

Now imagine an instrument whose polar axis does not lie in the meridian plane, but inclines toward the East (Figure 8-3). If we aim at a star in the direction of South, the point followed by the scope tends to rise with respect to the star's path. Through the telescope, you can see the star "ascend" the eyepiece field.

The second rule of the Bigourdan method is: if a star in the South ascends (or descends, respectively) the field of the eyepiece, it does so because the mounting inclines toward the East (or the West, respectively).

With successive sightings toward the South and East (or West) you can thus correct all alignment errors in the polar axis of an equatorial mounting. The precision of the method can be very great and depends upon the patience of the operator. You can consider the maximal duration of sighting a star to reach one hour. During this hour, a star near the equator travels 15°, or 54,000 seconds of arc. Now, with the help of a reticulated eyepiece, you can guide through a drift of a few seconds of arc: 3″ corresponding to 0.03mm at the focus of a 2m focal length instrument, or of 1m focal length equipped with a Barlow—a drift of 3″ at the end of an hour's tracking corresponds to a polar axis alignment error of 11″! The odds are that the axis' mechanical alignment will not let you exploit the full potential of this technique.

VIII.1.3. — THE PHOTOGRAPHIC METHOD

This method was developed in 1977 by the author. Its aim is to calculate the alignment error following measurements of blurring made upon a photograph. We represent the causes of blurring by three

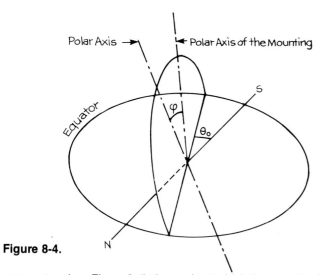

Figure 8-4.

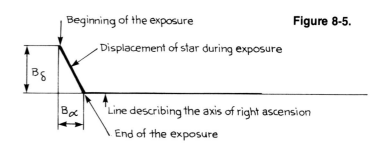

Figure 8-5.

where δ is the declination of the star photographed, Θ its hour angle, and where ω, the speed of the earth's rotation, must be expressed in the same units as the time t and the angles B α and B δ.

Since we need three equations to determine the three unknowns φ, Θ_o and $\frac{\Delta \omega}{\omega}$, we must photograph two stars of different declinations and hour angles. For precision in calculation, it is even advisable to photograph three stars in order to eliminate systematically the smallest values of B α and B δ.

This method suffers from the inconvenience of requiring precise focusing (you must find a star whose luminosity is just enough to expose the film without causing a large, unmeasurable spot) and a film development followed by a calculation. It is not more precise than Bigourdan's method, but it lets you determine the exact position of the polar axis in the space of a single operation and gives at once the drift in the drive. It can also represent a source of inspiration for all the devotees of astronomical manipulation and calculation. . . .

parameters (see Figure 8-4): the angle φ which the mounting's polar axis makes with the earth's polar axis (this angle is zero when the mounting is correctly oriented); the angle Θ_o which is the hour angle of the meridian plane containing the polar axis of the mounting; finally, the average relative drift $\Delta \omega / \omega$ of the drive system.

Unguiding photography of a star during time t reveals a blur: the star leaves a streak on the film. After exposing, if you cut off the drive motor without closing the camera shutter, the star's displacement, owing to the movement of the sky, will describe the axis of right ascension. You can thus measure the components B α and B δ of the blur according to the axes of right ascension and declination (Figure 8-5).

Now, we can express the values B α and B δ as a function of the exposure time t and the three parameters δ, Θ_o, $\frac{\Delta \omega}{\omega}$:

$$B \alpha = t \omega (\cos(\Theta - \Theta_o)\sin \delta + \frac{\Delta \omega}{\omega} \cos \delta)$$
$$B \delta = t \omega \sin (\Theta - \Theta_o),$$

VIII.2. — FOCUSING

VIII.2.1. — WHAT IS FOCUSING?

When we take a picture of a point object with a telescope (a star, for example) all the light coming from it is captured by the circular surface of the objective, then focused into a point by the divers intermediate optical systems. It is this point that we must make coincide with the film plane in order to obtain the point image of the star.

In order to simplify the diagrams and calculations, we shall consider that the object photographed always lies at the center of the field, that is, on the principal axis of all the systems involved. This restriction will not in fact have any notable influence. The bundle of light coming from a star, then, will always be symmetrical about the principal axis.

By way of example, let's take the case of a prime-focus photograph (Figure 8-6). The bundle of light entering the instrument takes the form of a cone whose base is the objective, whose height is the focal length of the latter, and whose tip lies at the instrument's focus.

Figure 8-6.

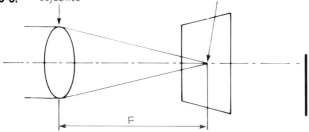

We can see that if the film does not coincide exactly with the focal plane (on the diagram below, it is too far back) the stellar image is no longer a point but a small disk: the picture is blurred (Figure 8-7).

Figure 8-7.

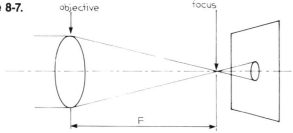

In general, whatever the optical system used, the bundle of light from a point object that reaches the film always takes the form of a cone. The film must always coincide with the tip of the cone.

This cone is characterized by the value of its half-angle at the tip. We shall call the "aperture" of the light bundle the tangent of this angle, that is, the ratio of the cone's diameter to the distance from the tip to the point where the diameter is measured (Figure 8-8).

Figure 8-8.

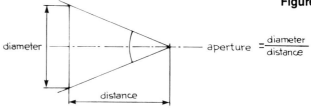

$$aperture = \frac{diameter}{distance}$$

For an instrument used at its focus, this aperture can be calculated instantly: it is the inverse of the ratio F/D (Figure 8-9).

Figure 8-9.

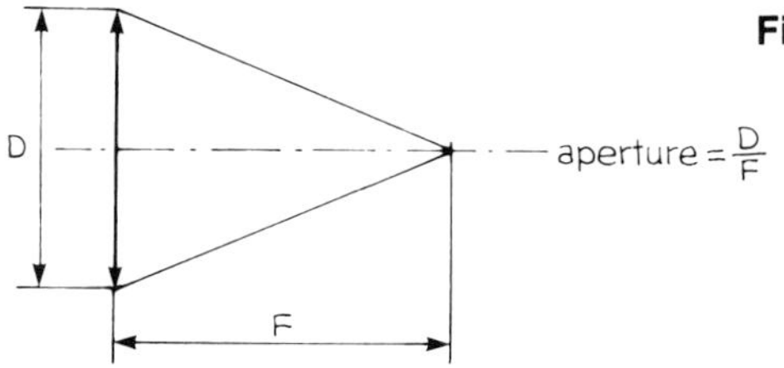

$$aperture = \frac{D}{F}$$

VIII.2.2. — THE IMPORTANCE OF FOCUSING

For all calculations of focus, we shall let the stellar image furnished by the system be a perfect point, that is, we shall let diffraction and the different aberrations remain negligible. This will not alter the results and has the merit of simplifying the reasoning.

Suppose, then, that the film plane does not exactly coincide with the point image (in reality, they never coincide perfectly). Let's call x the distance separating them. This focusing error is translated on the film by the representation of the star as a small disk of diameter ϵ (Figure 8-10).

Figure 8-10.

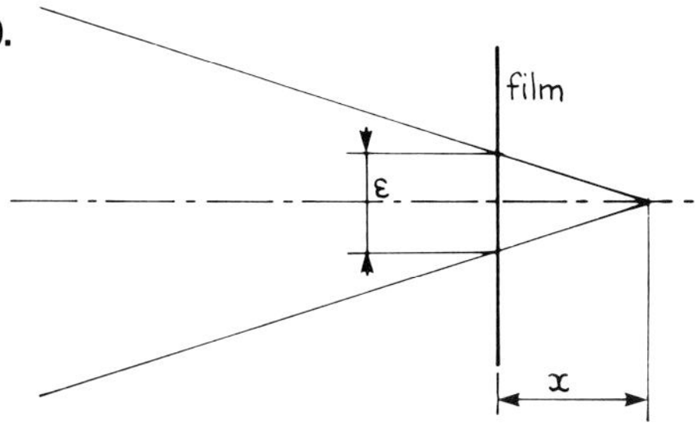

film

ϵ

x

The ratio ϵ/x equals the aperture of the light bundle. If, for example, the aperture is 0.1 (which is the case for a prime-focus picture with a scope of F/D = 10) x equals 10 ϵ. Now, if we require the disk representing the star to be no greater than the granulation of the film, which we shall fix at, say, 20 μm, this implies that x (that is, the precision in positioning the film) is less than 0.2mm!

This example shows what care you must bring to focusing. *Few amateur astronomers are aware of this, many of them looking for the finest film possible and not realizing that they are dividing the resolving power by 4 or 5 by poor focusing.*

Since in order to do serious astrophotography we must always make the finest possible focus, we shall first explain the calculation for determining the required level of precision, and then explore the different methods for achieving it in actual practice.

VIII.2.3. — ATTAINED PRECISION

In this section we give the formulas relating the diameter ϵ of the blurred spot on the film to the positioning error x of the focusing system for the setups most often used.

VIII.2.3.1. — PRIME-FOCUS PHOTOGRAPHY

This is the simplest case, since the aperture of the light bundle is by definition the inverse of the ratio F/D of the scope and since the focusing can be done only by moving the film holder.

Figure 8-11 shows the proportion relation:

$$\epsilon/x = D/F, \text{ or } \epsilon = x/(F/D), \text{ or } x = \epsilon \times F/D.$$

Figure 8-11.

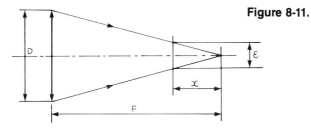

We see, then, that the positioning of the film will demand more precaution the smaller the focal ratio. For example, it will have to be three times more exact with a Newtonian telescope of F/D = 5 than with a refractor of F/D = 15.

VIII.2.3.2. — USE OF AN ENLARGING SYSTEM

In every case, the light emitted from the last optical element is focused on the film in the form of a cone whose aperture is defined by:

$$\text{aperture} = 1/\mu \times D/F_o = D/F_e,$$

where F_o is the scope's prime focus and F_e the effective focal length given by the magnification μ (Figure 8-12).

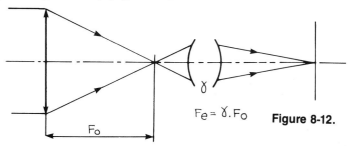

$F_e = \gamma \cdot F_o$

Figure 8-12.

Focusing is effected by the relative displacement with respect to the film of the final image obtained. This can be done in three different ways:

1) *Moving the film with respect to the telescope and the enlarging system, which remain fixed.* Thus, we are considering the same case as prime-focus photography, except that the focal length to account for is now the effective focal length:

$$x = \epsilon \times F_e/D.$$

Since the F_e/D ratios used in planetary photography are very large, a fairly small precision in positioning the film will do. For example, for $F_e/D = 100$, a maximal error of 1mm limits the blurring of the focus to $10\mu m$ on the photograph. On the other hand, you must prevent any great disturbance of the filmholder, which is why this system is the least used.

2) *Moving the enlarging system and the camera (interconnected) with respect to the telescope.* This is the case when the camera is attached to the movable portion of the eyepiece tube. We can here consider that it is the primary image of the telescope which moves a value of x with respect to the enlarging system of which it is the object. The image given by the enlarging system is thus displaced by $\mu^2 x$. This value is multiplied by the aperture of the light bundle at the film level in order to obtain the amount of blur:

$$\epsilon = D/F_e \times \mu^2 \times x,$$

from which we get:

$$x = \epsilon \times F_e/D \times 1/\mu^2,$$

and since $F_e = \mu F_o$,

$$x = \epsilon \times F_o/D \times 1/\mu.$$

Contrary to the previous setup, the adjustment for this one proves more difficult the greater the magnification used. Every astrophotographer has already noticed this phenomenon. For example, if the magnification is close to the primary F_o/D of the telescope (which is often the case by pure coincidence) then the precision in positioning the optical elements is of the same order as the acceptable value for blur, that is . . . a few tens of microns! Users of long focal lengths must realize one thing: *if you do not control the focusing mechanism perfectly, a large magnification will make you lose so much resolution in the absence of proper focus that you gain nothing by increasing the size of the image.* You will notice as well in the formulas given that for the same final F_e/D, the instruments with a large primary F_o/D are favored, since the magnification μ decreases as F_o increases.

The Schmidt-Cassegrain telescopes found commercially come with a device which focuses by moving the primary mirror along its axis, all the other optical elements remaining fixed. As far as the calculation of error is concerned, this system is analogous to the setup we have just seen. The Cassegrain mirror, however, must be accounted for in the enlarging system. For instance, for a photograph at the Cassegrain focus (with $F/D = 10$) you must apply the formula $x = \epsilon \times F_o/D \times 1/\mu$ with $F_o/D = 2.5$ (aperture of primary mirror) and $\mu = 4$ (magnification of secondary mirror), which hurts the precision required for the focusing mechanism.

3) *Moving the enlarging system alone, the camera body remaining fixed to the telescope.* Calculation shows that we need only replace μ^2 by $(\mu^2 - 1)$ in the formulas for the previous case:

$$x = \epsilon \times F_e/D \times 1/(\mu^2 - 1).$$

In general, μ is large enough for us to assimilate $(\mu^2 - 1)$ with μ^2, so that all the remarks pertaining to the previous case apply here also.

VIII.2.4. — FOCUSING METHODS

VIII.2.4.1. — GROUND GLASS

This method consists of replacing the film with a sheet of glass that is very finely ground (frosted) on one side and placed exactly in the film plane. (Thus, the same image forms on the ground glass as on the film (Figure 8-13). We can observe this image to determine its quality. Focusing consists of adjusting the mechanism until the image formed on the ground glass becomes sharp.

Ground glass focusing is the method most often used in amateur cameras; however, it would be difficult to replace the film with a ground glass for every picture. As a result, these cameras contain a mechanism, known as a reflex system, that uses a retractable mirror to direct the light

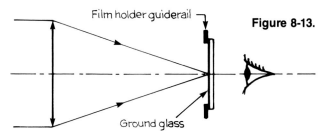

Figure 8-13.

toward the glass for focusing; then, the instant just before the picture is taken, the mirror is retracted and allows the light to reach the film (Figure 8-14). In this system, the light must cross exactly the same distance to reach the glass as it does to reach the film in order for the images formed to be identical. For finer focusing, one can put a loupe between the glass and the eye.

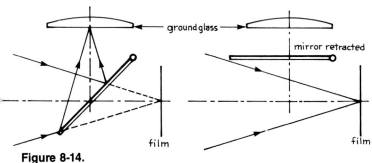

Figure 8-14.

This method of focusing is ill suited to the problems of astrophotography. The sharpness obtained on the glass usually falls short of astrophotography's demands. Above all, the loss in luminosity across the glass is such that the eye can not judge image sharpness with sufficient precision. This problem does not exist in ordinary photography, since the objects are much more luminous than those of interest to the astronomer.

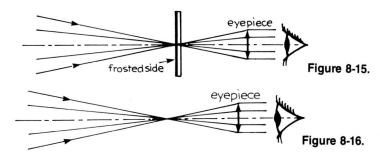

Figure 8-15.

Figure 8-16.

VIII.2.4.2. — REMOVING THE GROUND GLASS

Let's look at Figure 8-15 showing a groundglass focusing system where a loupe is placed in front of the eye. If we remove the ground glass, this setup now resembles an astronomical refractor, where the loupe plays the part of an eyepiece (Figure 8-16). The eye continues to see the image formed by the objective—it is found at the focus of the eyepiece lens when it appears sharp. We need only make the focus of this lens coincide with the film plane, and the ground glass is no longer necessary. We have at the same time eliminated the problem of lack of luminosity for judging the focus.

There is still another problem to tackle, however. Imagine that the image produced by the objective does not form completely at the loupe focus, and consequently, not entirely within the film plane. The picture will be blurred, but the eye potentially does not perceive this because its crystalline lens accommodates itself automatically to compensate for this small defect. The process must thus be improved. With this goal in mind, we place in the film plane one side of a clear sheet of glass, furnished with an engraved reticle (Figure 8-17).

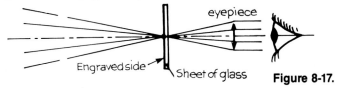

Figure 8-17.

If the eye perceives the telescopic image and the reticle simultaneously, it is because it has accommodated simultaneously to them both; thus, they lie in the same plane. The focusing procedure consists of first adjustment the loupe-reticle distance so that the reticle appears sharp—and is thus close to the loupe focus. Next, we focus the telescopic image until it is sharp in turn. Be careful: when the eye fixes upon the telescopic image while the adjustment is off, it can accommodate itself to this image, which will appear sharp without the observer's realizing that the reticle has become blurred. You must strive to bring them both into focus simultaneously.

If you are focusing on the moon, the reticle appears in sillouette on our satellite's surface. On the other hand, if you are focusing on a star, the reticle is usually invisible against the dark background of the sky. You should therefore illuminate the telescope objective slightly in order to perceive star and reticle simultaneously against the slightly luminous background sky.

Experience shows that the point of focus for a star procures the greatest focusing precision. The chosen star must appear sufficiently bright and brightness depends upon the focal ratio used. But too bright a star can be a problem since turbulence easily surrounds it with crests that can conceal a defective focus.

One variant of this method sometimes allows for greater precision. It consists of placing the star inside a line on the reticle and moving the in front of the lens, perpendicularly to this line (Figure 8-18). If the star actually lies within the reticle plane, it will seem fixed. Conversely, if it lies farther back than the eye, it will seem to move in the same direction as the eye with respect to the reticle, and in the opposite direction if it is closer.

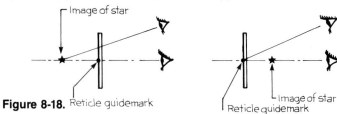

Figure 8-18. Reticle guidemark

Image of star
Reticle guidemark

In practice, then, what material should you use? A preliminary solution consists of reticulating the front of a sheet of glass and fixing an eyepiece to the back, taking care to make the eyepiece focus coincide with the reticulation (Figure 8-19). You need then only apply this assembly to the film holder guide rails.

Figure 8-19.

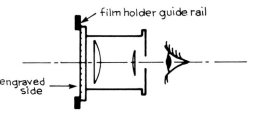

film holder guide rail

engraved side

It is simpler, however, to use a reflex camera whose ground glass can be replaced by a clear, reticulated glass. You must then put a focusing lens behind the eyehole, or, which is better when the camera permits, replace the traditional viewfinder by a viewfinder loupe.

VIII.2.4.3. — FOUCAULT'S METHOD

Here we must place in the film plane a sheet of opaque material (a razor blade, for example) which must intercept the light rays in part of the field. The eye observes the edge of the blade, usually without intermediate optics.

Let's aim at a star. Without the blade, the eye beyond the focus receives a diverging bundle of light (Figure 8-20). Since each of the rays making up this bundle comes from a point on the objective, the eye sees the objective disk lit up by the star.

Figure 8-20.

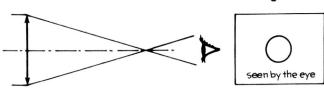

seen by the eye

Now let's place the blade perpendicular to the axis of this bundle so that it intercepts it (Figure 8-21). If the blade lies in the same plane as the point of focus, it intercepts the entire bundle simultaneously at this point. The eye sees the objective extinguished altogether.

Figure 8-21.

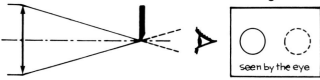

seen by the eye

On the other hand, if the blade is closer to the objective, it cuts off the rays issuing from the same same side as the blade (Figure 8-22). The eye thus sees the objective partially extinguished and the shadow progressing in the same direction as the blade.

Figure 8-22.

seen by the eye

Conversely, if the blade lies farther from the objective than the point of focus, it intercepts the rays coming from the opposite side—the shadow moves in the reverse direction (Figure 8-23).

Figure 8-23.

seen by the eye

With this method, then, you can ferret out the point of focus, that is, the telescopic image in the plane of the blade, which is also in the film plane. The method is very precise but not always very simple to put into practice. It is not recommended for instruments which tend to vibrate. Photographers who can adapt a powerful enough loupe to their cameras often prefer it over this method.

In practice, it is hard to construct a setup using a moveable blade in the film plane. It is easier to attach the blade to the film holder guide rails and move the star with respect to it by adjusting the slow motion controls of the instrument.

VIII.2.4.4. — STAR TRAILS

This method consists of photographing a star field with a stationary exposure. The stars leave trails on the film, owing to the apparent motion of the sky. The width of each trail depends upon the star's luminosity, but also upon the precision of focus. You have to make a series of pictures of the same field, slightly altering the focus each time, within a range certain to contain the correct setting. The developed pictures are next examined with the enlarger or under a microscope. Choose from the field a dim star, so that its trail will not be enlarged by the diffusion phenomena of the light in the gelatin. The picture on which the trail of the star is the finest is the closest to the correct point of focus.

This method is perhaps the most precise of all, but it has a few inconveniences. First of all, it is cumbersome and costly: you must sacrifice a certain number of photos, develop them, and examine them very closely. The number of photos you have to take can be very great if you don't have precise enough an idea what the correct setting is; thus, you must narrow down the range using one of the other methods. Next, the setting must be reproducible, that is, the system must be furnished with guide marks precise enough to relocate the position it was in when the best test photo was taken. Finally, the star trail method has the inconvenience of locating the focus *a posteriori*. Now, the focus can vary with time, especially because of heat expansion in the optical or mechanical pieces. It can happen that the setting which seems the best on the test photos will not be correct after the time it takes to develop the film.

You can improve this method by making an experimental correction meter for the ambient temperature. You will still have to measure your adjustments with great precision and in a reproducible fashion for several nights, and you must have the patience to apply the star trail method for a certain number of different temperatures.

VIII.2.5. — FOCUSING AND FINENESS OF FILM

At this point in our presentation, it is evident that the image sharpness is limited, owing to focusing errors, by two factors: the mechanical sensitivity of the focusing system and the sensitivity of the method for judging this focus. It is useless to use a film which can reach an image fineness unattainable at the point of focus. This seems obvious, but very few amateur astronomers take it into account. Consequently, you must adapt your focusing system to the performance of the film used, or if this is not possible, use a coarser film that will in compensation enjoy the advantage of greater sensitivity.

VIII.2.6. — FOCUSING AND RESOLVING POWER

Obtaining the instrument's theoretical resolving power in a photograph can be considered as the goal to achieve in astrophotography. The spot of diameter ϵ into which each elementary point of the image is transformed because of a bad focus must be less than the corresponding value for the resolving power.

We know that the resolving power for a telescope in seconds of arc is 120/D, where D is the objective diameter in millimeters. R.P. = k/D radians, where k is a constant (k = 0.58×10^{-3}). If we use an effective focal length $F_e = \mu F_o$ (where μ is the magnification of the enlarging system and F_o the prime focus of the objective) the distance separating two details whose spacing corresponds to the resolving power is k/D $\times \mu F_o$.

If we require that the precision of focus correspond to the instrument's resolving power, we write ϵ = k/D $\times \mu F_o$.

Now, we know that ϵ depends upon the setting precision x of the focusing system. We have seen that in most cases we can allow x = F_o/D $\times 1/\mu \times \epsilon$.

From this formula and the preceding expression for ϵ we determine the setting precision necessary to attain the instrument's resolving power to be x = $k(F_o/D)^2$ where (k = 0.58×10^{-3}).

For example, with a telescope whose primary mirror is open to F_o/D = 10, the movable piece (eyepiece, Barlow, or film) doing the focusing must be positioned at close to 0.058mm (58 microns!). This example dramatically demonstrates that a picture's fineness is quite often limited by the focusing precision rather than by the instrument's actual capabilities.

Let's examine the previous expression more closely: x = $k(F_o/D)^2$.

Remarkably, the magnification does not enter here. The greater the magnification, the more sensitive the adjustment; however, the larger the image of an object, the less it needs to be magnified. These two effects compensate for each other exactly. Consequently, the necessary mechanical precision with the focusing system remains constant, whether we are photographing at the focus or using any sort of enlarging system.

The second thing to notice in the formula is the influence of the focal ratio, F_o/D. You might be surprised to learn that it is not the effective focal length which is involved but the primary focal length F_o of the objective. As the ratio F_o/D increases, so does the tolerance x for the focus. Since F_o/D in the expression is squared, its influence becomes great: we have to be 9 times more exact in positioning the focusing system with a scope at F_o/D = 5 than with a scope at F_o/D = 15.

VIII.3. — SPECTROGRAPHY

Spectrography consists of taking pictures of the luminous spectrum of a body. This discipline demands specialized equipment, but lies within the reach of the amateur astronomer. At this level, however, the only objects of interest are the sun, stars, and nebulae.

VIII.3.1. — THE DIFFRACTION GRATING

The grating is a sheet of glass etched with a large number of close spaced parallel lines. When a bundle of parallel light rays reaches it, part of the light crosses through without being changed (order 0) while the rest are deflected and dispersed according to their wavelength (orders 1,2, etc. see Figure 8-24). The light rays are deflected in a plane perpendicular to the lines on the grating at an angle Θ which depends upon the wavelength:

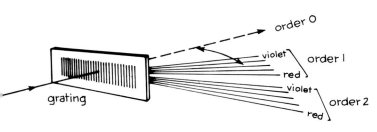

Figure 8-24.

$$\sin \Theta = nk\lambda \ ,$$

here λ is the wavelength of the light ray in millimeters, n the number of ~~nes~~ per millimeter on the grating, and k a whole number. For k = 0, the ~~undle~~ of light is neither deflected nor dispersed (order 0). For k = 1, 2, ~~.,~~ we obtain different deviations corresponding to spectra of orders 1, 2, ~~. .~~

III.3.2. — STELLAR SPECTROGRAPHY

III.3.2.1. — SETUP

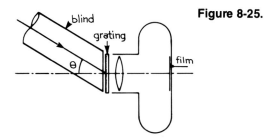

Figure 8-25.

Figure 8-25 shows a grating mounted in front of a camera lens ~~et~~ at infinity and with a blind in front of the grating so that it is illuminated ~~nly~~ by the stars lying in a direction corresponding to the average devia- ~~on~~ of the first spectral order (we can work just as well with the second order;

the luminosity is weaker, but the dispersion twice as great). For example, for a grating of 600 lines per millimeter, the visible light undergoes a deviation between $\Theta = 14°$ for $\lambda = 400$nm and $\Theta = 25°$ for $\lambda = 700$nm. A star is thus represented on the film by as many point images as there are wavelengths in its spectrum. In fact, since the spectrum is continuous, the star is represented by a line each point of which corresponds to a wavelength, from violet at one end to red at the other.

But a spectrum in the form of a very thin line would not be very intelligible. We must give it a certain height, which we do by slowly displacing the star in a direction parallel to the lines on the grating and thus perpendicular to the spectral dispersion. We then obtain a rectangular image where each wavelength is represented by a line (Figure 8-26).

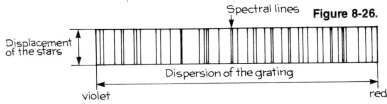

Figure 8-26.

In practice, the easiest method is to utilize the sky's apparent motion to displace the star. Since this displacement is too rapid for the time needed to expose the film, however, we let the star move for a time Δt corresponding to a chosen height for the spectrum on the film (1mm works well), and then replace the star at its point of departure at the end of this interval Δt. We repeat the operation until the end of the entire exposure. Of course, we must take care to place the lines of the grating parallel to the celestial equator so that the star's displacement becomes perpendicular to the dispersion of the grating.

Successive attempts should be made with the help of a guide scope furnished with a recticulated eyepiece. If this guide scope must also be used for aiming at the chosen region of the sky, you must remember to incline the dispersion angle Θ of the grating with respect to the camera axis.

Table 8-1 gives for two commercial gratings and for different camera focal lengths the length L of the spectrum for wavelengths between 400 and 700 nanometers and the displacement time Δt corresponding to a height of 1mm on the negative for a star close to the celestial equator (in the other cases, multiply δt from the table by cos δ , where δ is the star's declination):

TABLE 8-1

Focal Length (mm)	Displacement Time t (min)	Spectrum Length (mm) 600 lines/mm	1200 lines/mm
35	6.5	6.8	13.5
50	4.6	9.7	19.3
85	2.7	16.4	32.9
135	1.7	26.1	52.2

The choice of film follows the same criteria as for faint-object photography. You must also make sure that the film selected covers the spectral region you want to photograph. There is no problem with 2415 and most common films. On the other hand, in the series of films corrected for reciprocity failure, you must select 103aF.

Photograph 8-1. *These two stellar spectra (Vega above, Sirius below) were made by Jacques Silvain by placing a grating (600 lines/mm) in front of a 135mm telephoto. The hydrogen lines are easily distinguished.*

Exposure time depends upon the star's magnitude, the film, the objective, and the dispersion of the filter, but also upon the spectral region. In practice, you must use several different exposure times in order to expose correctly all the zones of the spectrum—a few dozen minutes is good.

VIII.3.2.2. — USING A TELESCOPE

The grating must receive parallel rays of light so it is placed directly in front of the camera lens. There should be no optics in front of the grating.

The setup we have just described presents an inconvenience, though: most gratings are small, which limit the light-collecting surface. Large gratings are very expensive. One solution consists of having a telescope serve as light collector.

We know that when a telescope is set at infinity, all the light received from a star leaves the eyepiece as a bundle of parallel light rays. This is the setting used for visual observation. The diameter of this bundle is the exit pupil d of the instrument, which depends upon the diameter D of the objective and of the magnifying power M used: d = D/M (see Section I.4.5.2.).

To increase the amount of light received by the grating, we place in front of the previous setup a telescope whose exit pupil is less than the dimensions of the grating (Figure 8-27).

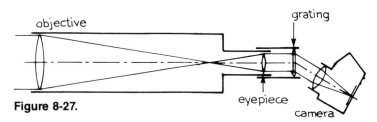

Figure 8-27.

You must make certain that the telescope is set perfectly at infinity. The actual process of photography remains the same as for the preceding setup. But, beware of the fact that the speed of the sky's apparent motion is multiplied by the magnifying power of the scope—you must thus divide the recommended values for Δt by M.

VIII.3.3. — SOLAR SPECTROGRAPHY

The sun, unlike stars, presents an apparent surface. If we use the previous methods in solar spectrography, each wavelength is no longer represented on the film by a point or a line, but by a disk (the image of the sun). Under these conditions the solar spectrum would become unusable because of the overlapping of these infinite disks. We must thus replace the circular image of the sun by that of a line. We can achieve this only by using a telescope.

Let's go back to the setup in which the grating is fixed between the camera and a telescope set at infinity. At the telescope focus an image forms of the sun. If we place a diaphragm in the form of a very thin slit at the image level, the grating and the film no longer receive the entire image of the solar disk, but only a ray of light (Figure 9-28).

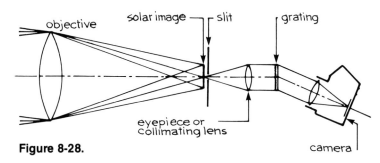

Figure 8-28.

The slit must lie parallel to the lines on the grating so that they are perpendicular to the grating's dispersion. The solar spectrum thus consists of the juxtaposition of an infinite number of images from the slit, each

Photograph 8-2. Christian Buil took this photograph of the spectrum of stars from the constellation Lyra. He placed a prism in front of an objective with a focal length of 400mm. The over exposed star is Vega.

corresponding to a given wavelength. The resolution $d\lambda$ in wavelengths of the spectrum is thus limited by the width ϵ of the slit. A simple calculation shows that

$$d\lambda = \cos\Theta/nk \times \epsilon/f_c,$$

where f_c is the focal length of the eyepiece.

We can see the need to make the slits as narrow as possible (juxtaposing two razor blades, for example) and in using an eyepiece of long focal length. Preferably one should replace the traditional eyepiece with a collimating lens, greater in focal length and whose focus, where the slit must be installed, is more accessible.

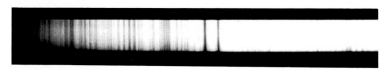

Photograph 8-3. *Philippe Alexandre and Michel Esteves took this picture of the lines of the solar spectrum. They made a spectroscope with a 600 lines/mm grating, an objective of 100mm focal length and a slit formed by two razor blades.*

Diffraction phenomena in the grating can also limit resolution. If N is the total number of lines in the grating, the resolution at wavelength λ can not pass the value $d\lambda$ such that $d\lambda/x = 1/N$. We can see the need for using a grating with as many lines as possible and for illuminating its entire surface. The setups used by amateurs are usually not precise enough, however, and the limitation of resolution comes rather from the size of the slit than from diffraction in the grating.

As in solar photography, you can use high-resolution, slow films and short exposures. The latter will depend upon all the factors already mentioned, to which is added the width of the slit. Each operator, then, will have to determine the suitable exposure time according to the particular setup.

VIII.3.4. — SPECTROGRAPHY OF NEBULAE

Like the sun, nebulae present an apparent surface; however, they differ in that they have a peculiar spectrum consisting almost entirely of emission lines. The sun and stars, whose surface temperatures are high, give a continuous spectrum, owing to thermal radiation, upon which absorption or emission lines superimpose themselves. However, nebulae are gaseous masses cold enough for thermal radiation to be negligeable in the visible range. On the other hand, the atoms composing them, excited by mutual collisions, emit light within certain well defined wavelengths. By way of example, Figure 8-29 represents the spectrum of the nebula M42 in Orion. The height of the lines indicates the relative intensity of the light rays. The diagram also specifies the ions or the atoms corresponding to each wavelength.

As a result, the spectrum of a nebula is not continuous, but discrete. We can thus proceed as we did for stars, without using a slit: the photograph of the spectrum of a nebula shows, then, a finite number of juxtaposed images of the nebula, each image corresponding to an emission wavelength of the spectrum.

The comparison of images of different wavelengths can inform us about the distribution of different gases inside the nebula. In order not to distort these images, the telescope must track the sky's apparent motion precisely (as in faint-object photography) whereas stellar spectrography requires a little sweeping.

VIII.4. — LABORATORY TECHNIQUES

VIII.4.1. — FILM DEVELOPMENT

If most amateurs prefer buying films commercially and having them developed, they must nevertheless realize that it is far better to learn to develop them themselves—not because the people in the photo industry are incompetent (quite the opposite) but because you will be asking them to do a task that does not correspond at all to what they are used to. Examples abound: a standard development which will do nothing to bring out astronomical images, standard printing with a 2mm planet in an 8 × 12 picture, film returned as unexposed (there was nothing but a few points

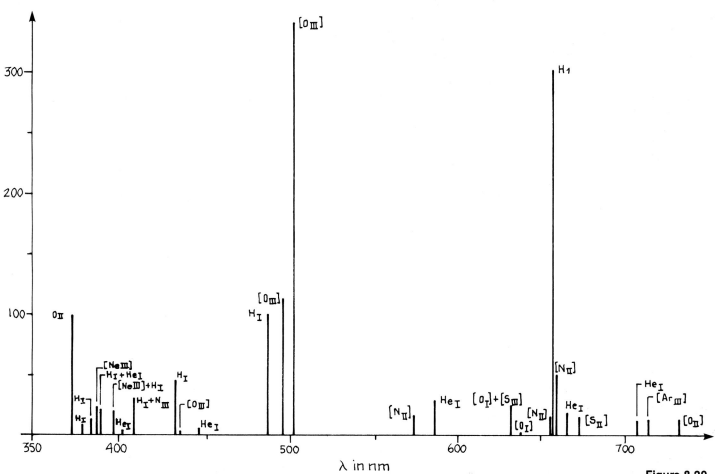

Figure 8-29.

on it!), color printing with a red or green sky according to the inspiration of the moment. If you spend the money to have an operational telescope suitable for astrophotography, it's only logical to make the effort required to learn how to develop your photos yourself and exploit your equipment's capabilities to the limit. In the following discussion, we shall assume that you know how to develop a standard black-and-white film, either because you have read an elementary book on photography, or because your associates have taught you. We shall content ourselves with providing advice specific to astrophotography.

1) *Work at a constant temperature: 20°C.* From developer to rinse, you must keep an eye on the temperature. The temperature of the developer is particularly important and you should avoid the general practice of developing for a shorter time because the developer is warmer. In fact, the developer does not act at all the same on the photographic grain at 15° as at 25°C. You must therefore work at 20°C. Likewise, everything else must be at the same temperature in order not to cause reticulation. If the film gelatin undergoes too many large temperature changes (for instance, developer-stop-fixer at 20°C, then tap water rinse at 10°C) it splits—and you wind up with a crackled photograph. Although artists sometimes seek this effect, astrophotographers had better protect themselves from it, since taking a two-hour exposure of a galaxy in order to then reticulate it during development does not make a whole lot of sense.

2) *Agitate strongly during development.* In order to obtain images of greatest contrast, you must agitate the tank at the time of development. Normal photographs accommodate themselves to a weak contrast (standard agitation during the first 15 seconds followed by 5 seconds every minute) but in astrophotography, you must agitate vigorously. You can rotate the spindle or turn the tank, one effective method being to combine the two.

3) *Use only fresh products.* If you want to obtain maximum results, you must work with good quality solutions. As far as developer goes, we advise you to use it only once. The common practice of using the same developer several times, increasing developing time in order to compensate for the loss in strength, should be avoided. It is better for you to use the exact quantity necessary (300ml for 35mm films) and to toss it out after use. After a vigorous agitation, the developer has oxidized, which degrades its performance for reuse. You can be less strict with stop bath and fixer.

You should, however, make sure they will keep well. You can buy special "light-tight" containers, as well as accordian bottles that let you keep developers away from air (you compress the bottle gradually as you use the product). Read and follow the manufacturer usage instructions for fixing and washing.

4) *Don't stray from the beaten path.* For example, you should be wary of "pushed" developments: you can increase the film's density, but not necessarily without diminishing the contrast, increasing the grain, or failing to keep clear of the background fog density. If you do wish to experiment, keep detailed notes so that you can replicate desirable results later.

VIII.4.2. — MASKING AND UNSHARP MASKING

We have already had occasion to point out the problem posed by the large range in luminosity of certain objects, a problem aggravated by the use of high-contrast films: certain zones are overexposed, whereas other, darker zones of the same object are underexposed. In planetary photography, this holds true for the moon (seas darker than mountainous regions and, especially, a terminator darker than the limb) for Venus and Jupiter (edge darker than the center) and for Saturn (ring darker than the planet).

We have advocated films with limited contrast so that the least luminous regions do not reach the saturation point of the characteristic curve for the emulsion, all the while allowing the darker zones to lie above the threshold of sensitivity. This is a necessary condition if you want to register details in both bright and dark regions.

But our concern for bringing out the least contrasty planetary details forced us to use the highest contrast film possible, out of respect for the preceding rule. As a result, every shade of gray in the emulsion represents a portion of the image. A problem arises, then, at the printing stage, since photographic paper does not have as wide a dynamic range as the negative. For example, Kodak 2415 developed in D76 represents the maximal admissible value of contrast for a quarter moon. A correctly exposed negative reveals details at the limb and at the terminator simultaneously. But upon printing, even on the lowest contrast paper possible, either the limb area remains a uniform white (for a short exposure on the paper) or the terminator disappears into blackness (for a longer exposure). We must therefore resort to masking.

Since the enlarging paper must be exposed longer in certain areas than in others, we leave it under the enlarger for the longest time required by the bright zones of the object, while masking the dark zones for part of the exposure. In the case of the moon, it is rather easy to match the shape of the terminator or the seas with your hands. You must place your hands the proper distance from the enlarger and move them continuously in order to reduce the amount of light.

You can also use a cardboard mask cut in the desired shape. For Jupiter, for instance, a piece of cardboard in the shape of a circular crown proves the most practical for masking the edges of the planet.

But the appropriate shapes for masking are not always so simple. For example, the nebula M42 shows volutes of very different luminosities. The center is more brilliant, but it is not defined by a simple geometric form. The best solution consists of using a picture of the nebula as a mask.

From the original negative (that you photograph) either by contact or by a slide reproduction system, you can obtain an image of the same scale, but this time in positive. If you superimpose the original negative with the positive in the enlarger at the moment of printing, the image obtained is a uniform gray, the two films producing opposite filtrations. If on the other hand you take care to underexpose the positive film, it will no longer annul the negative image; rather, it only reduces its contrast. Note, though, that it reduces the microcontrast of fine details you wish to preserve as well as the troublesome macrocontrast between zones.

The solution lies in using as a positive an image of the same scale as the negative, underexposed, but blurred. Thus, the luminous intensities are averaged out among the large areas and the masking changes nothing of the contrast between neighboring details; on the other hand, it diminishes the global contrast from one region to another. You can obtain this blurred positive by interposing a diffusing surface at the time of duplication, or by making an imperfect focus during this operation, but you must be careful to maintain a uniform magnification. The most rigorous solution is to obtain the positive by duplicating a second photograph of the object taken at the telescope for which you have purposely degraded the focus.

VIII.4.3. — REVERSAL

Contrary to the case we have just examined, a negative can suffer from too little contrast. The appearance of Kodak 2415 has allowed us to avoid this problem in planetary photography. In faint-object photography, on the other hand, too faint an object can turn up on the negative scarcely any darker than the background fog—we must thus augment the picture's contrast before printing it on the paper.

Toward this end we make the original negative with a high-contrast film. The latter must also have a very fine grain or be exposed with the help of an enlarger (we proceed as for a normal print, the reversal film simply replacing the enlarging paper) in order not to alter the fineness of the first image. We obtain a new image, higher in contrast, but positive—we must repeat the maneuver once more to obtain a negative ready to print and giving us the desired contrast.

This operation proves highly effective for revealing faint galaxies or the tails of comets.

VIII.4.4. — COMPOSITES

After three-color printing (Section VII.4.2) and masking (Section VIII.4.2) here is the third special printing technique used in astrophotography.

The aim of compositing is to reduce the influence of the film's grain. Instead of taking one picture of a celestial object, let's take a large number of identical pictures (under the same conditions and in an interval of a few minutes only, if it is a matter of a planet whose aspect changes). From this series of negatives we select 4 to 10 of the best quality. Rather than print them separately, we can expose them successively on the same paper, taking care of course to center them in exactly the same way and to divide the total exposure time needed for the paper among all the negatives.

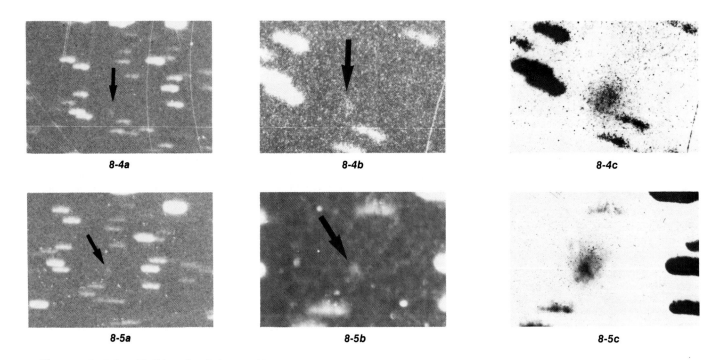

8-4a 8-4b 8-4c

8-5a 8-5b 8-5c

Photographs 8-4 and 5. This series of photographic prints shows the power of duplicating photographic negatives onto high contrast lithographic film to detect faint objects. Eric Laffont and the Author made two exposures of Comet Halley on February 19, 1985 when it was between 19 and 20 magnitude. This appears to be the first unambiguous photograph to be taken by amateur astronomers of the comet's 1985/86 return. The photographs were made at Pic du Midi using a 60 cm reflector with a focal length of 2.1 meters on Kodak 2415 hypersensitized film and exposed for one hour each.

The first print in each series is of the original negative. These prints were then successively re-copied on Kodalith film and printed to enhance contrast. Note the overall condition of both the "a" and "b" prints. These pictures were taken at 9,000 feet with an ambient temperature of −14°F. The photographers thought that the film had warmed sufficiently for processing but in reality it had not. When emersed in developer at 70°F the emulsion deformed (reticulated). Between prints "b" and "c" most of the effects of this deformation were removed by opaquing them out on the negative. All prints were done on variable contrast paper (grade 5). Notice that while the comet is barely visible in print "a" it becomes progressively more visible in "b" and "c".

Wherein lies the interest of this operation? If the successive negatives are well centered, the same details of the object correspond to the same points on the paper. On the other hand, the film grain, which distorts these details, presents with respect to them a chance distribution independent from one negative to the other. Statistics teaches us that a chance uncertainty averaged over n measurements is divided by a factor of $\sqrt{n}$. In our case, the composite print from n negatives will present an apparent granulation $\sqrt{n}$ times smaller. For example, we register a gain of 2 for a composite of 4 negatives, a gain of 3 for 9 negatives, etc.

Photograph 8.6. Dominique Albanese first made three good exposures of Saturn with a 203mm telescope. Then in the darkroom he made a composite print from these three negatives which noticeably reduced the grain (Ilford HP5).

Composites yield spectacular effects with films having a strong granulation. Some amateurs tend to consider it as the miracle solution. We should mention a few reservations, however.

You must note well that compositing is useful *only* when the image quality is limited by the film's resolution and in no other instance. This happens when the mediocre reliability of a mounting requires a short exposure and thus a focal length too small with respect to the film's resolution. Never can compositing improve a photograph if its quality is limited by turbulence. In fact, the latter would introduce from one picture to the next apparent chance displacements in details greater than those you might dread during exposure.

In our opinion, the appearance of Kodak 2415 has rung the death knell for compositing insofar as the extreme fineness of this film means that picture quality is usually not limited by the film grain, but by focusing, turbulence, or the resolving power of the instrument.

Nevertheless, compositing would still be useful with the 103a films, which have a large grain, and thus will faint-object photography. But it is difficult to expect an operator to make a 30-minute track of the same object ten times! In such a case, using a more powerful, hypersensitized film seems the better solution.

VIII.5. — SATELLITES

It is possible to photograph the satellites of Jupiter and Saturn and even—for the better equipped—those of Uranus and Neptune. The most brilliant of these appear like fifth-magnitude stars, but by definition they lie quite close to a brilliant body, which presents an additional problem.

In fact, photographing a satellite requires exposures of several seconds or several minutes on a fast film—you can calculate the time needed with the formula from Section II.7. During this long exposure, however, the planet (which lies in the field) so saturates the film that it leaves a spot much more extended than the actual size of its image. Outside of a dubious aesthetic effect, the result is that the satellite's image tends to drown in this blotch and in every case prevents a precise measurement of the satellite's position with respect to the planet. The difference in illumination proves much too great to be compensated for by masking during printing.

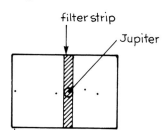

Figure 8-30. Example of placing a filter so that it masks Jupiter yet passes light from the Jovian satellites.

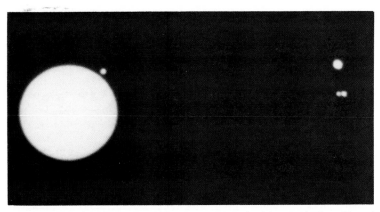

Photograph 8-7. *Jean Dragesco took this photograph of the occultation of beta 1 and 2 Scorpionis (double star of second magnitude) by Jupiter and its satellite Io. This is a rare event — occurring once every 4 or 5 centuries. Here we see the emersion of beta 2 Sco on 5/14/71. This observation lead to the conclusion that Io had no atmosphere.*

We must therefore mask the planet while shooting, which is a much more difficult operation. The best solution consists of using, just before the film, a strip of gelatin filter whose width slightly exceeds that of the planet (Figure 8-30). The filter's coefficient of transmission depends upon both the planet and the desired exposure time (thus upon the magnitude of the satellite that you wish to attain); it is generally in the neighborhood of 1%.

The greatest difficulty lies in centering the planet on the filter. First of all you must establish the filter strip's position with respect to a reference point in the viewfinder. You can do this by photographing a stationary target with this system in place. Of course, you must then aim with sufficient precision and ensure good tracking during the exposure.

CHAPTER IX
EQUIPMENT

IX.1. — THE TELESCOPE

IX.1.1. — OBJECTIVE DIAMETER

We know that the image quality provided by an instrument improves as the diameter of its objective increases. The telescope's theoretical resolving power is the smallest apparent angle one can hope to resolve under ideal conditions. Expressed in seconds of arc, its value is R.P. = 120/D, where D is the objective diameter in millimeters.

Naturally, in astrophotography as in visual observation, we want to use an instrument with the largest possible diameter. Yet we can accomplish successful astrophotography with the most modest of instruments; the subjects available are simply more limited. Below we give the minimum diameters for attempting certain subjects:

- 60mm: moon, sun, photos with camera used "piggyback";
- 80mm: Jupiter, rings of Saturn;
- 100mm: Mars, Venus, Saturn, satellites of Jupiter;
- 150mm: satellites of Saturn.

We give these values only as an indication, since an instrument's capacities depend a lot upon the optical quality, the mounting, and the site.

Performance in faint-object photography depends as much upon guiding as upon the diameter and proper guiding requires additional equipment. This is why we omitted this area from our table above.

IX.1.2. — REFRACTORS OR REFLECTORS?

We would scarcely be exaggerating if we said that the small world of observing is divided into two clans: those who favor refractors and those who favor reflectors.

We know that the theoretical resolving power of an instrument depends only upon its objective diameter—it thus remains independent of the technology used and must be identical for a refractor and a reflector of the same diameter. Nevertheless, partisans of refractors attribute to them a superior image quality. How so?

IX.1.2.1. — OBSTRUCTION OF THE SECONDARY MIRROR

A reflector, whether Newtonian or Cassegrain, always has a secondary mirror, placed in front of the primary mirror (Figure 9-1). We shall refer to the ratio d/D of the diameter of the small mirror to the diameter of the primary mirror as the obstruction. As a rule of thumb, this obstruction takes on values between 0.18 and 0.40. For instance, a 100mm scope with a secondary mirror having a diameter (or a small axis if it is an elliptical Newtonian mirror at a 45° angle) of 25mm, gives an obstruction of 0.25.

Figure 9-1.

The consequence of an obstruction is loss of light, since the secondary mirror casts a shadow on the primary mirror. This loss is in fact rather limited and has practically no effect on image quality. Indeed, in percentage it is equal to the ratio of the surfaces of the two mirrors and thus equal to the square of their diameters. An obstruction of 0.3 means that the surface covered by the secondary mirror is 0.09 times the surface of the primary mirror. Thus, it causes a 9% loss in the total amount of light that the latter can collect, which is hardly catastrophic: from the sole point of view of luminosity, everything occurs as if we were using an unobstructed scope whose diameter is decreased by 4%.

The main problem lies elsewhere: the presence of the secondary mirror in front of the objective alters the diffraction image given by the latter, and its influence becomes greater the greater the obstruction. In the absence of all obstruction and for perfect optics, we know that the image of a star is a small disk surrounded by fainter rings (see Section I.4.7).

A secondary mirror reinforces the size and luminosity of the first ring of the diffraction disk. For an obstruction less than 0.2, this phenomenon remains scarcely detectable, but for an obstruction of 0.3 or more, we must practically consider the useable image of the star to be a small spot no longer limited by the diameter of the central disk, but by the external diameter of the first ring. The resolving power is not excessively degraded if you want to split stars of identical luminosity, but the reinforcement of the first ring becomes much more troublesome for star pairs having a large magnitude difference, or planetary images, which are low in contrast. In this case, we can consider that for a large obstruction (that is, close to 0.3 or greater) the telescope's resolving power is altered by a factor of about 2. From the point of view of resolution, everything occurs as if we were using an unobstructed instrument whose diameter is no more than .55 times the actual diameter. This explains why images always appear less sharp and lower in contrast in a reflector with too much obstruction.

IX.1.2.2. — TUBE CLOSURE

In refractors the objective is situated at the entrance of the main tube. In reflectors, however, the objective consists of the parabolic mirror at the base of the tube. The top of the objective tube is usually open and thus comes in contact with the surrounding air. If thermal equilibrium is not perfectly maintained, air currents arise, transforming the tube into a chimney and creating very strong turbulence. Obviously, such a turbulence in the tube particularly mitigates against image quality.

Figure 9-2a.　　　　　　　**Figure 9-2b.**

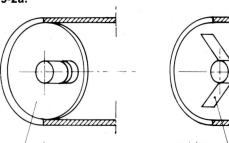

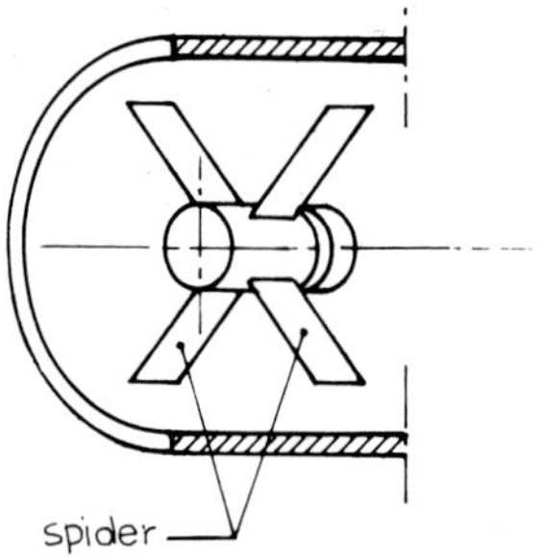

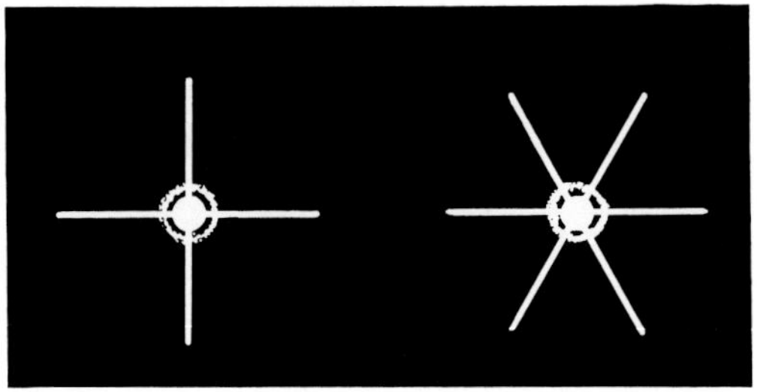

window　　　　spider

Figure 9-3. A "spider" with four veins causes four spikes to appear on the diffraction disk of a star: likewise, a spider with three veins gives rise to six diffraction spikes.

The best remedy for eliminating this turbulence consists of covering the tube entrance with a plane parallel glass window which must be made with the precision demanded of a refractor objective (Figure 9-2a).

The covering window enjoys another advantage: it is usually pierced through the center, so that you can use it to attach the secondary mirror support. Normally, this support is attached to the tube by an assembly of three or four blades called a "spider" (Figure 9-2b). The spider, lying in the path of the light rays captured by the objective, causes spikes to appear in the diffraction disk (Figure 9-3). These spikes abet the degradation of the image provided by the reflector. The main inconvenience of a window is its high price: a good one usually costs more than the primary mirror. If anti-reflective coated the light loss for most applications is minimal.

IX.1.2.3. — CHROMATIC ABERRATION

We know that a lens deflects light rays of one wavelength differently than those of another wavelength. To correct this phenomenon, refractors come with an objective with two lenses, called achromatic. The chromatic correction made by such an objective is not perfect, however, and the position of the focus varies slightly according to the wavelength considered. Although of lesser importance than obstruction or tube turbulence (which are the failings of reflectors) chromatic aberration, which

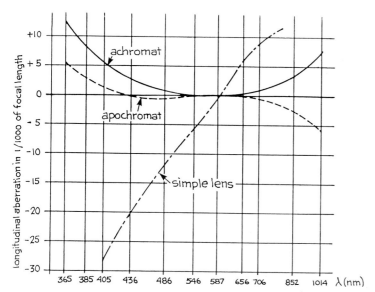

Figure 9-4. *This graph represents the variations in focal length as a function of wavelength for a simple single element lens, an achromatic objective with a 2 element lenses, and an apochromatic objective with a 3 element lenses.*

Photograph 9-1. *The Author took this photograph of the nebula M42 at the Newtonian focus of the 60 cm telescope at the Pic du Midi Observatory, F/D = 3.5 on 103aF film exposed for 3 minutes. This instrument's large aperture causes noticeable coma effects which are clearly visible at the edges where the stars are radially elongated.*

occurs only with refractors, can alter images slightly, since a focus rigorously made for one color is incorrect for the others.

A better correction for aberration comes from objectives with three lenses, known as "apochromatic," which are unfortunately much more expensive. The image furnished by a reflector is entirely free from chromatic aberration, since the light does not cross any piece of glass at the objective level.

IX.1.2.4. — FOCAL RATIO

An instrument's luminosity is related to its focal ratio F/D (objective focal length divided by its diameter). For prime-focus photography of faint objects, we want the ratio as small as possible; however, we must take into account certain constraints.

The faster a parabolic mirror is (that is, the smaller F/D is) the more the images present aberrations known as "coma" as soon as one leaves the principal axis. The images of stars are no longer points, but are lengthened radially with respect to the center of the field. These mirrors should therefore not fall below F/D = 5. In a Newtonian scope, the obstruction of the secondary mirror increases as F/D decreases (the necessity of having a secondary mirror of large diameter to intercept the entire bundle of light without causing vignetting). Newtonians usually have focal ratios between 5 and 10. For the same reasons, the obstruction of a Cassegrain varies substantially as the inverse of the magnification u of the secondary mirror. This magnification is the ratio between the effective focal length F_e of the Cassegrain system and the prime focus F_o of the parabolic mirror.

To limit obstruction, μ would have to be as large as possible, which would lead to resulting F_e/D ratios too large for faint objects, since $F_e/D = \mu \times F_o/D$. The search for the best quality of image for a Cassegrain would require at least $F_o/D = 5$ and $\mu = 5$, which gives us a final $F_e/D = 25$. In order not to hamper the observation of faint objects, commercial instruments do not respect these constraints, and one finds Cassegrains from $F_e/D = 10$.

Refractors are particularly limited by chromatic aberration, which increases as F/D decreases. Scopes exist with ratios between 10 and 20.

IX.1.2.5. — IN CONCLUSION

The differences between reflectors and refractors that we have just explained show that, for equal diameters, a refractor (if possible, with large focal ratio) is always preferable for high resolution photography. However (diameters still being equal) refractors cost much more than reflectors. For equal price, you can usually obtain a reflector twice the diameter of a refractor and thus having twice the theoretical resolving power. We should therefore ask ourselves which instrument is better adapted to the various types of astrophotography for a predetermined price.

IX.1.2.5.1. — Solar photography

Resolution is at least as limited by turbulence problems as by the instrument's diameter. The impossibility of maintaining thermal equilibrium, owing to the intense heating caused by sunlight, militates against open tubes. Furthermore, the ability to select relatively narrow spectral bands, with the aid of filters, permits us to suppress chromatic aberration. The refractor, therefore, proves indispensable for solar photography.

IX.1.2.5.2. — Planetary photography

Tube turbulence can be better combatted here than for the sun, and a reflector, even with an open tube, can prove very effective, provided that its primary mirror is not too fast and that the obstruction is reasonable. For planetary photography experience will not let us easily decide between a refractor and a reflector at the same price, that is, of an appreciably superior diameter.

IX.1.2.5.3. — Faint-object photography

Our main problem here is lack of light; thus, the instrument's diameter becomes practically the sole criterion. For the same price, the choice belongs without hesitation to a reflector, especially a Newtonian. Reflectors and Cassegrains, though, are handicapped by high focal ratios (F/D $\leq$ 10) which require prohibitive exposure times for faint objects. While some manufacturers offer focal reducers, they are not, however, a miracle solution, (loss of light through the crossing of supplemental lenses, limited field, sometimes insufficient optical quality).

IX.1.3. — THE MOUNTING

In astrophotography the quality of the mounting becomes at least as important as the quality of the optics. Too many amateurs forget this fact. Indeed, the sharpness of the images obtained is often limited not by the resolving power of the objective, but by a blur for which the mounting is responsible. We saw in Chapter V that to counteract this, we must limit the exposure time according to the mounting's reliability (vibrations, tracking). Now, a reduction in exposure time implies a reduction in focal length (in order to satisfy the luminosity equation) and thus in a reduction in the size of the image obtained.

IX.1.3.1. — ALTAZIMUTH MOUNTINGS

Because of the necessary combination of their two movements, these do not permit sufficiently reliable tracking for photos requiring high resolution. Moreover, the only subject compatible with a tracking conceivable for altazimuth mountings is wide-field stellar photography— but the long exposures this requires prove incompatible with the problem of field rotation (see Section I.3.4.1) related to altazimuth mountings.

The only cases remaining for altazimuth mountings, then, are those for which the exposure is short enough not to require tracking. This applies to the sun and, in a pinch, to the full moon, at the instrument focus.

In these cases, the altazimuth mounting serves solely for aiming. You must, therefore, keep it free of large vibrations. Recall that objects at the celestial equator move $1''$ in a fifteenth of a second of time. Instruments mounted in altazimuth usually have a small diameter; one can not expect from them a resolution better than 2″. To aspire to this, however, you must not take exposures longer than 1/30th sec., which corresponds to an object displacement of 0.5″ at the equator.

IX.1.3.2. — NONMOTORIZED EQUATORIALS

The only important advantage in photography here compared to altazimuth mountings is the suppression of field rotation. You can thus look forward to long exposures in faint-object photography. Since tracking is done manually, the attained resolution depends upon its regularity. Of course, this resolution corresponds to the maximum focal length possible for the photographic objective.

We can estimate that the sudden jerks caused by manual tracking correspond to small stops lasting up to one second before correction is completed, and then only if the operator has good training. This is the time needed to detect a drift and react to it, or to reposition the hand on the control when required. The adjustment made during this second of interruption must not exceed $30\,\mu m$, which closely represents the resolving power of a fast film. This limit fixes a maximum useable focal length of 400mm for an object at the equator, or 800mm for a declination of 60°.

In actual fact, we do not recommend faint-object photography at the telescope focus (using, for example, a guide scope at the edge of the field) for nonmotorized mountings. Only piggyback setups, perhaps, prove feasible with an objective of moderate focal length.

Some amateurs train themselves to do regular, well-gauged tracking for two or three seconds. Without the efficiency of a motor, this procedure allows them to limit the influence of the sky's rotation. Thus, the maximum admissible exposure for a high-resolution photo can be multiplied by 4 thanks to this technique. The limit we have set at 1/30th sec., in the case of stationary shots for altazimuth mountings, here becomes 1/8th sec. This gain lets us try prime-focus lunar photography outside the period of the full moon, but remains insufficient for other areas of planetary photography.

IX.1.3.3. — MOTORIZED EQUATORIALS

These prove indispensible in most cases. They must not generate vibrations, must be easily adjustable so that the polar axis can be aligned with the North Celestial Pole with the greatest precision, and must have regular, precise gears and motors. Manual adjustment in right ascension and declination must be possible during long-exposure photography. We have covered all these characteristics of motorized equatorials in the chapters devoted to planetary and faint-object photography.

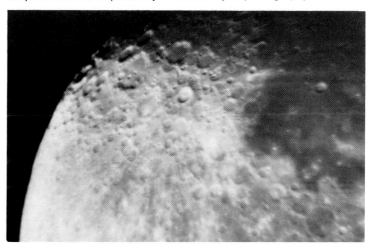

Photograph 9-2. The Author took this photograph using a 115mm telescope and a Huygens eyepiece as an enlarging system to demonstrate that when focused on the center field (region of Clavius) the edge of the field is seriously out-of-focus.

IX.2. — EYEPIECES AND BARLOWS

It does you no good to have a perfect objective if the optics of your enlarging system are of poor quality.

Orthoscopic or Plossl eyepieces prove the best adapted to high-resolution photography, especially if the telescope's primary focal ratio is small.

Simpler eyepieces, such as the Huygens, usually can not focus simultaneously for the center and the edge of the field. While unimportant as far as planets are concerned, this defect poses a problem for lunar or solar photography.

In solar photography you must watch out for overheating at the eyepiece. Orthoscopic or Plossl eyepieces, as well as Barlow lenses, contain achromatic doublets in which the two lenses are glued together with Canada balsam. If the sunlight does not travel through some sort of filter or helioscope before reaching the eyepiece, overheating can alter the balsam and unglue the lenses or they may crack due to differential expansion!

IX.3. — THE CAMERA

A camera well-suited for astrophotography has two important features: it permits precise focusing using one of the methods described in Section VIII.2, and it does not generate vibrations at the moment of shooting.

The most popular astrophotography camera is the 35mm reflex. We shall therefore refer especially to these, though the general principles apply to all other cameras.

We should mention that most specialized astrophotography films (Kodak 2415, 103a, etc.) are available or easily attainable only in 35mm format. The sole inconvenience of this format is that it imposes a maximum focal length of 2.40m for photos of the moon or sun in their entirety.

IX.3.1. — FOCUSING

We have seen that the standard ground glass does not suit astrophotography. We must therefore be able to replace it by a clear, reticulated glass. We must also replace a pentaprism viewfinder by loupe or be able to adapt a powerful loupe behind it. (Figure 9-5) Obviously, the reticulated face of the glass must lie at an optical distance from the objective rigorously equal to that of the film.

When the reflex setup can not be changed, or when it is absent, we can make a system with a loupe and a sheet of glass engraved on the front face, adaptable to the film holder guide rails before the film is put in. Another solution consists of using Foucault's method (see Section VIII.2.4.2.

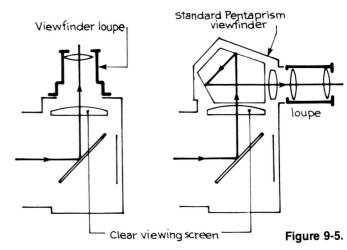

Figure 9-5.

and VIII.2.4.2.) but here again you suffer the inconvenience of not being able to focus with the camera already loaded.

IX.3.2. — VIBRATIONS

These arise at the moment of shooting, caused by the movement of the shutter curtain and especially by the lifting of the reflex mirror. It becomes important, therefore, to have the power to retract the mirror manually a few seconds before shooting, thus suppressing the main cause of vibration. Unfortunately, cameras equipped with a manual retraction button are rare.

Lifting the mirror in advance, however, entails its own inconvenience: as soon as it is done, you can no longer see the image of the object you want to photograph. You can not, therefore, inspect the change in turbulence in order to shoot during a moment of calm. As a result, manual retraction is not used systematically. The choice depends on the one hand upon the rigidity of the setup (telescope, mounting, camera) and on the other hand upon the amount of turbulence.

In every case, shooting must be done as smoothly as possible. Manufacturers seem to have made a move in this direction with modern cameras. In the never-ending fight against vibrations, you must always use a cable release rather than pressing the button with your finger.

IX.3.3. — AUTOMATION

Automatic cameras set the exposure time themselves, which they calculate from the film speed and a measure of luminosity provided by a cell. Such automation does not suit astrophotography. Indeed, a cell which integrates the luminosity over a large portion of the field—and thus over the dark background sky surrounding, for example, a tiny planetary image—tends to overexpose. The camera must therefore have a manual setting so that the operator can achieve the exposure time calculated beforehand.

IX.3.4. — SOME RECOMMENDED CAMERAS

We shall limit ourselves to 35mm cameras, which are the most widespread and the easiest to use. It is in this format as well that you can find the special astronomy films Kodak 2415 and 103a.

A few years ago, Miranda made cameras with an interchangeable viewfinder and viewing lens, and a manually retractable mirror. Accessories included a clear, reticulated viewing glass, a loupe of small magnifying power, and a viewfinder magnifying 5 and 15 times. Thus equipped, the Miranda G and the Miranda Laborec were the ideal cameras for astrophotography. Unfortunately, Miranda went bankrupt in 1976 and these cameras are now very hard to find second hand.

The only cameras with interchangeable viewfinders now on the market are the so-called professional cameras, whose main point in common is a prohibitive price: the Pentax LX, the Nikon F3 (replacing the F2 and before that the F) and the new Canon F1 (succeeding the old F1). The LX, however, has no viewfinder of magnifying power beyond 1.5x, which will not suffice to achieve a correct focusing. The Canon does not have a manually retractable mirror. Owners of the LX can adapt an 8 to 12mm focal length eyepiece in place of the viewfinder. This eyepiece replaces an enlarging viewfinder without any problem, but nevertheless you may need to fiddle

Photograph 9-3. A Miranda G equipped with its 5× and 15× viewfinder. Also shown are interchangeable loupe and a clear reticulated glass.

with it to make its front focus coincide exactly with the reticle. You must thus choose an eyepiece whose focus lies as far ahead of the field lens (orthoscopic or Plossl) as possible. This arrangement—using an eyepiece—is not possible if you can not remove the standard pentaprism from the camera, since you can not bring the eyepiece close enough to the viewing lens.

Among the cameras without removable viewfinders, the Olympus OM1 appears the most attractive: a smooth release, a manually retractable mirror, the ability to adapt to a clear viewing glass and a loupe magnifying 2.5x behind the pentaprism. A 2.5x magnifying power is still too weak to control focusing correctly for small focal ratios. You can therefore construct a powerful loupe with two lenses. Indeed, the viewfinder loupes of cameras are nothing more than tiny telescopes set at infinity. Their magnifying power is the ratio of the focal length F of the objective lens to the focal length f of the ocular lens (Figure 9-6).

IX.4. — VIGNETTING

Vignetting is the loss of light at the edge of the field, related to the optical assembly. Newtonian and Cassegrain reflectors are particularly subject to vignetting, the extent of which depends upon how they are designed. In addition, the optical assemblies used in photography (Barlows

Figure 9-6.

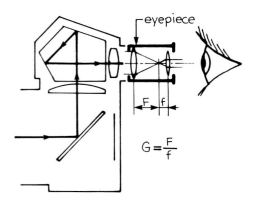

$$G = \frac{F}{f}$$

in particular) can introduce vignetting. You will benefit, then, by knowing the limits of your equipment in this area.

XI.4.1. — DESCRIPTION OF VIGNETTING

Imagine a simple optical assembly: an objective of radius R and focal length F and a diaphragm of radius R/2 placed midway between the objective and its focus (Figure 9-7).

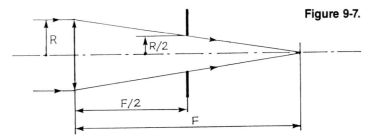

Figure 9-7.

In this setup all light rays penetrating the objective and concentrating at the focus pass through the hole in the diaphragm. For a star at infinity lying on the objective's principal axis, the diaphragm has no effect.

Photograph 9-4. *P. Le Fur took this photograph of the galaxy M33 at the focus of a 200mm Schmidt-cassegrain telescope equipped with a focal reducer (F/D = 5) on 103aF film exposed for 45 minutes. The background sky has fogged the center of the picture, delimiting the fully illuminated field. Around this area the phenomenon of vignetting is illustrated — at the edges of the field, which are less illuminated, only the most brilliant stars have exposed the film.*

Consider now a star situated off the principal axis such that its image forms in the focal plane at a distance R from the axis. Figure 9-8 illustrates that only the light rays passing through the bottom of the objective reach the focal plane; the others are blocked by the diaphragm. For the star in question, this phenomenon results in a loss of luminosity, about half of what the objective could supply in the present case. The loss of light is usually nil near the center of the field, but increases the farther away we move. This phenomenon is called *vignetting*.

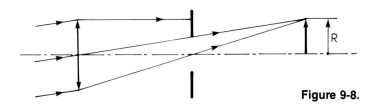

Figure 9-8.

Camera lenses or enlarging objectives usually have a certain number of diaphragms the aim of which is to improve image quality by eliminating aberrant marginal light rays. But these diaphragms cause vignetting. This fact is well known by photographers and is often measured when photographic magazines test lenses.

On astronomical instruments, the causes of vignetting are not numerous, but you should know them and above all know how to anticipate their consequences.

IX.4.2. — THE NEWTONIAN TELESCOPE

In this common scope, the light collected by the objective is sent toward the eyepiece tube by a mirror inclined 45° (Figure 9-9). S represents the length the bundle of light is bent. The diagonal mirror is elliptical so that, seen at a 45° angle, it seems circular, from the eyepiece tube as well as from the objective.

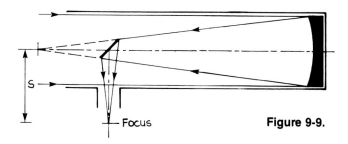

Focus   Figure 9-9.

The light rays reflected by the primary mirror which pass beside the secondary mirror are lost to observation. We can thus consider the secondary mirror as a diaphragm and replace the study of the Newtonian scope by that of the simpler arrangement shown in Figure 9-10.

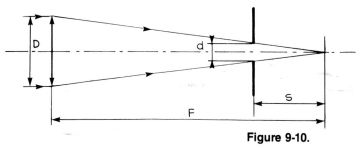

Figure 9-10.

To simplify even more, we can assume this diaphragm to be circular, perpendicular to the principal axis and equal in diameter to the small axis of the secondary mirror (Figure 9-11).

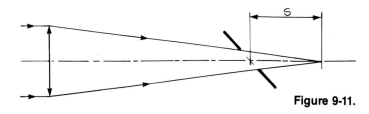

Figure 9-11.

In order for all the light collected by the objective to reach the center of the field (at the focus) the secondary mirror must satisfy the relation

$$d/S > D/F.$$

The obstruction created by the secondary mirror (that is, d/D) must thus satisfy

$$d/D > S/F.$$

Now, S is at least equal to the radius of the primary mirror. In practice, the focus must lie sufficiently beyond the tube to permit the assembly of optical systems. Thus, S is on the average close to the diameter of the primary mirror for a small amateur scope. If we suppose S = D, the obstruction becomes

$$d/D > 1/(F/D).$$

Notice that the obstruction is related to the focal ratio. The smaller this ratio, the greater the obstruction becomes. For example, if F/D = 5, to have full light at the focus requires $d/D \cong 0.2$. Now, we know that 0.2 is the limit we must observe if we wish to maintain the scope's theoretical resolving power.

Figure 9-12.

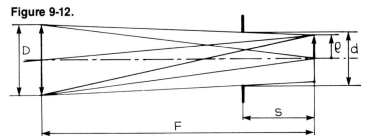

Of course, the condition imposed upon the small axis d of the secondary mirror becomes even more constrained if we wish to keep all the light collected by the objective over an extended field (Figure 9-12). Let ρ be the radius of the field of full light. The geometric construction of the light path permits us to write

$$d = DS/F + 2\rho (1 - S/F).$$

The term $(1 - S/F)$ is usually close to 1; hence:

$$d/D = S/F + 2\rho/D.$$

Most often the amateur astronomer wants to know the field of full light given by a telescope. We can determine this by letting ρ be the radius of this field and S the distance separating the focus from the principal axis of the primary mirror:

$$\rho \cong \frac{1}{2}(d - S/(F/D)).$$

Outside this field, vignetting occurs.

Let's finish our discussion of Newtonian scopes by considering the example of the two Vixen 100mm diameter scopes now on the market. The 100/1000, with a 1000mm focal length, has a secondary mirror with a 22mm small axis, and its focus extends with respect to the axis of the main tube S = 220mm. This telescope has a reasonable obstruction of 0.22, which allows very good images. Its focus, however, is much too extended and the ratio d/S is just equal to D/F: full light is thus available only at the center of the field. At 10mm from the center, we have already lost half the luminosity to vignetting. This instrument, well designed for the high resolution of a small object (planet) does not suit wide-field photography.

The Vixen 100/600, on the other hand, with a 600mm focal length, has a large obstruction of 0.4 (small axis of secondary mirror: 40mm) which greatly hampers high resolution. Nevertheless, despite a value of S = 180mm, which is already high for this type of scope, it conserves the full light over a field of 10mm radius.

IX.4.3. — CASSEGRAIN TELESCOPES

In a Cassegrain scope the light no longer travels to the side but to the rear of the tube. The primary mirror, therefore, contains a hole to let the light pass behind the tube.

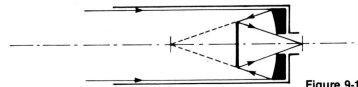

Figure 9-13.

Imagine that we were still using a flat secondary mirror. As Figure 9-13 shows, the light would have to be reflected at least halfway between the mirror and the focus in order for the latter to reach the end of the tube. We can also see from the figure that in order to collect all the light from the objective, the flat mirror must have a diameter equal to half the objective's—hence, an unacceptable obstruction of 0.5.

To avoid this problem, the Cassegrain system uses a convex, hence divergent, secondary mirror.

As you can see in Figure 9-14, the convex mirror can lie close to the primary focus and yet cast the secondary focus (that is, the focal point after both mirrors have acted) far enough to reach beyond the end of the tube. This secondary Cassegrain mirror must be hyperbolic, hence the term "hyperbolic mirror."

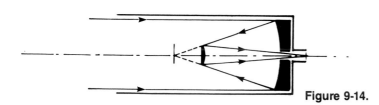

Figure 9-14.

Using a hyperbolic mirror, however, does affect the optics. It introduces a magnification μ into the system. The effective focal length F_e of the assembly equals μ times the focal length F_o of the primary mirror, and the size of an object at the secondary focus equals μ times the size of the object at the primary focus:

$$F_e = \mu F_o.$$

Let ρ and ρ' be the distances from the primary focus and the secondary focus to the hyperbolic mirror (which are the positions of the object and the image with regard to the principal planes, defined in the chapter on optics). Thus, the magnification μ equals the ratio of these distances:

$$\mu = \rho'/\rho.$$

As with the Newtonian telescope, we can dimension the secondary mirror so that it intercepts all the light concentrated by the objective into its focus. The condition is written (see Figure 9-15):

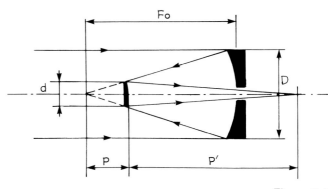

Figure 9-15.

$$d/\rho \geq D/F_o,$$

which gives us an obstruction

$$obs = d/D \geq \rho/F_o.$$

If we note that F_o lies near ρ' in this type of scope, we see that $obs = 1/\mu$. Now, recall that the primary and effective focal ratios are related, like the focal length, by the relation $F_e/D = \mu F_o/D$, which we can write in the first approximation as $F_e/D = 1/obs \times F_o/D$.

Here we have put our finger on the problem with Cassegrain telescopes: a small obstruction and an F_o/D sufficient to guarantee a certain optical quality imply a high final F_e/D. (For example, if obs = 0.2 and F_o/D = 6, then F_e/D = 30). Now, we know that the higher F_e/D becomes, the less luminous an image we get. With a telescope open to F_e/D = 30, exposure times are 25 times as long as with a scope open to 6. The Cassegrain, then, proves ill suited for faint objects.

Up to now we have required full light only at the center of the field. If we wish to benefit from an extended field with no loss in luminosity over the entire surface, the hyperbolic mirror must be larger. If the radius of the useful field at the Cassegrain focus is ρ, it corresponds to a field of radius ρ' at the prime focus with, of course, the relation $\rho = \mu\rho'$ (Figure 9-16).

Figure 9-16.

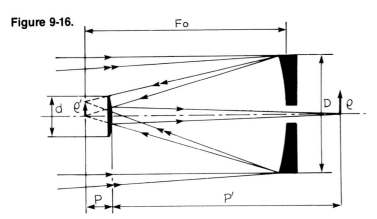

As with the Newtonian focus, we have the relation

$$d = D\rho/F_o + 2\rho' (1 - \rho/F_o),$$

which we approximate by

$$d/D = \rho/F_o + 2\rho'/D \cong 1/\mu (1 + 2\rho/D.$$

We see that the obstruction will increase for a small magnification μ and a large desired field at the Cassegrain focus. We can also use this formula to calculate the radius ρ of the field of full light, knowing the characteristics of the instrument.

For sometime, three companies—Celestron, Dynamax, and Meade—have been selling compact Schmidt-Cassegrain telescopes. These are in fact Cassegrain scopes whose primary mirrors are not parabolic, but spherical, the spherical aberration being remedied by a correcting plate. The interest in this design (spherical mirror + correcting plate) lies in preserving an image of uniform quality over an extended field. These Schmidt-Cassegrain scopes are sold as all-purpose scopes, adapted at the same time to high resolution and faint objects. This latter area, however, imposes upon the manufacturers an effective F_o/D limited to 10. Now, we have seen that this ratio is related to the obstruction and to the primary F_o/D. Consequently, Schmidt-Cassegrain scopes have primary ratios that are too small ($F_o/D \cong 2.5$) and obstructions that are too large (0.3 to 0.38) to provide images with enough contrast for high resolution. In fact, the resolving power of these scopes for objects low in contrast is that of a perfect instrument of half the diameter.

IX.4.4. — THE BARLOW LENS

You will recall that in photography we are sometimes driven to using a Barlow lens to obtain an effective focal length double the primary focal length of the objective.

The setup, then, is as shown in Figure 9-17. To obtain a magnification of 2, the image formed by the objective (hence the prime focus) must lie halfway between the Barlow lens and its front focus—thus, at f/2 from the lens, if f is its focal length.

Figure 9-17

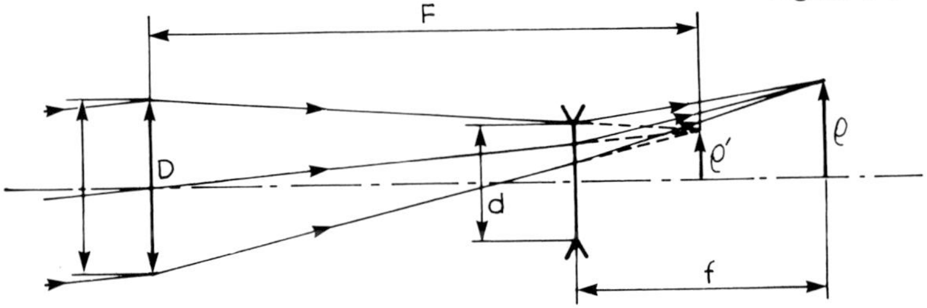

The Barlow has a limited diameter d; thus, it behaves like a diaphragm. Analysis of the diagram above shows that no light loss occurs at the center of the field if

$$d/(f/2) > D/F \text{ or } f/d < 2F/D,$$

F and D being the focal length and diameter of the objective.

In general, we can obtain this condition without any trouble. For example, for a Clavé Barlow, f = 114mm, d = 25mm; thus, f/d = 4.6. We can use this lens for the center of the field on every instrument open to F/D > 2.3, which is always the case.

On the other hand, it is interesting to calculate the field of full light obtained with a Barlow. Let ρ be its radius; it corresponds to a field of radius ρ' at the instrument's prime focus, and since the Barlow doubles the focal length, $\rho = 2\rho'$.

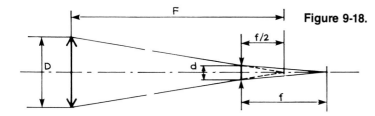

Figure 9-18.

Considering that the focal length of the Barlow is small compared to that of the objective, the geometric construction in Figure 9-18 gives us the relation

$$d = D/F \times f/2 + 2\rho',$$

hence,

$$\rho = 2\rho' = d - f/(2F/D).$$

If we reconsider the example of the Clavé lens previous, mounted on a telescope open to F/D = 5.7, we obtain $\rho = 15$mm. The diameter of the field of full light, 30mm, is greater than the diameter of the Barlow lens! This is due to the latter's diverging effect.

APPENDIX A

REFLECTIONS UPON LIMITING MAGNITUDE IN STELLAR PHOTOGRAPHY

We have shown in Section II.7 the following empirical law allowing us to calculate the attained limiting magnitude in stellar photography:

$$M = 8.4 + 5\log D + 2\log T - \log F \text{ (for ASA 800).}$$

If we analyze this law, it expresses that the amount of light received per unit of surface on the film is proportional to $D^2T^{0.8}/F^{0.4}$. The corresponding law for objects of nonzero apparent surface involves the term D^2T/F^2. Where do these differences come from?

The term D^2 expresses the amount of light collected by the objective. If the image were really a point, the surface luminosity would be directly related to the total luminosity and thus proportional to D^2T only (F would not come into it). If, on the contrary, we suppose that the stellar image on the film presents a spreading caused by diffraction of the objective (perfect optics), the surface luminosity must be proportional to D^4T/F^2. In fact, the angle of diffraction varies as $1/D$ (see Section I.4.7), the size of the image of diffraction on the film varies as F/D, thus its surface as $(F/D)^2$, and the surface luminosity is therefore proportional to $D^2T \times (D/F)^2$.

Comparison with the empirical law's $D^2T^{0.8}/F^{0.4}$ shows that the stellar image formed at full aperture is supposed to be independent of diffraction and thus determined by the quality of the optics. The slight dependence upon F ($F^{0.4}$) lets us suppose that all commercial objectives (whatever their focal length) are designed to have nearly the same resolution at full aperture, expressed in image size on the film, of the object resolved (and not in apparent angle resolved) with, however, a slight reduction in quality when choosing extenders of ever-increasing power. Notice that these reflections are entirely congruent with the tests which photographers are used to, where the optical quality is expressed as the number of lines resolved per millimeter, measured on the film.

The fact that the luminosity at which a film is sensitive varies as $T^{0.8}$ instead of being proportional to T simply expresses reciprocity failure (see Section IV.3.1).

APPENDIX B

EVALUATION OF SURFACE LUMINOSITY

In order to determine correctly the parameters which must satisfy the luminosity equation, we must evaluate the amount of light per unit of apparent surface sent to us by the body photographed. In practice, we shall seek to determine this surface luminosity not absolutely, but relative to a body for which we know the correct parameters. For example, if object A has a surface luminosity double that of object B, for which we already know the right exposure conditions, we need only choose for A an exposure twice as short as for B, everything else being equal.

Generally, we know the object's apparent dimensions and its magnitude m. The total amount of light we receive from it can be expressed as 2.5^{-m}, by definition of the magnitude. It is thus easy to compare the surface luminosities L_A and L_B of the two objects A and B for which we know the magnitudes m_A and m_B and the apparent diameters d_A and d_B: For example, at their 1982 opposition the apparent diameters of Mars and Jupiter were 14.8″ and 44.5″, respectively; thus, their magnitudes were −1.2 and −2.0. Hence:

$$L_{Mars}/L_{Jupiter} = 2.5^{(-2.0 + 1.2)} / (14.8/44.5)^2 = 4.3.$$

Mars is, per unit of apparent surface, 4.3 times as luminous as Jupiter.

Another calculation can be made for planets. The light they receive from the sun is inversely proportional to the square of the distance D separating them. The amount of light they return equals the amount received multiplied by their coefficient of reflection, known as *albedo*. The luminosity in apparent surface does not depend upon the planet's distance from the earth. In fact, the total amount of light received and the planet's apparent surface are inversely proportional to the square of this distance at the same time—the two cancel each other. If D_A and D_B are the distances from the sun of two planets A and B, the ratio of their apparent surface luminosities is thus:

$$L_A/L_B = (D_B/D_A)^2 \times \text{albedo A} / \text{albedo B}.$$

The Table B-1 gives for each planet the average distance from the sun, the albedo, and the average apparent surface luminosity calculated as above, taking that of Jupiter as unity.

TABLE B-1

	D (A.U.)	Albedo	L
Mercure	0.387	0.055	24.0
Venus	0.723	0.64	80.0
Moon	1.0	0.146*	9.4*
Mars	1.524	0.154	4.3
Jupiter	5.203	0.42	1.0
Saturne	9.555	0.45	0.32
Uranus	19.218	0.46	0.08
Neptune	30.110	0.53	0.038

*for the full moon

You must remember, however, that we define the concepts of albedo and magnitude for a spectral response identical to the eye's. The spectral sensitivity of films can deviate greatly from this and vary from one film to another. For example, Mars, more red overall than Jupiter, will seem to have a greater difference in luminosity with respect to Jupiter on a film with extended sensitivity in the red, such as Kodak 2415, than on a panchromatic film, such as Ilford Pan F.

APPENDIX C
EVALUATING TURBULENCE

You will recall that the perfect theoretical image of a star provided by a telescope is a small disk surrounded by diffraction rings (see Section I.4.7). The radius of this small disk in seconds of arc is a = 140/D, where D is the objective diameter in millimeters. By observing the alteration undergone by the diffraction image, we can evaluate with respect to "a" the amplitude of the spreading due to turbulence. We reproduce here a table given by Danjon and Couder in *Lunettes et télescopes*.

t < a/4: perfect images, without detectible distortion and scarcely agitated

t = a/4: entire rings, pervaded by mobile condensations

t = a/2: moderate agitation, broken diffraction rings, central spot has wavy edges

t = a : strong agitation, rings faint or absent

t = 3a/2: image tending toward a planetary aspect

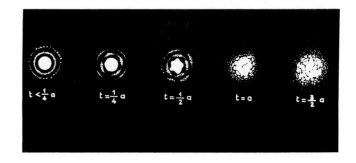

One method of evaluating the erratic fluctuations of images, linked to atmospheric agitation, consists of photographing a star using a very long focal length. The equatorial motor having been shut off, the star leaves a trail owing to the sky's apparent motion. The image of the star's trajectory is not a perfectly straight line, but a line distorted by fluctuations related to turbulence. Beware: do not confuse a periodic deformation in the trail caused by instrument vibration with the chance granular deformations of turbulence. A focal length of 10m allows you to have on film a scale of 50 microns for one second of arc; on an emulsion like 2415, this proves quite sufficient for getting a good picture of the turbulence. You must choose a star with a luminosity just sufficient to produce a thin line and the focus must be exact. If the star is a double, the spacing between the component stars can furnish a ready made scale upon the photograph.

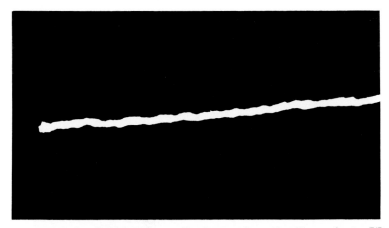

Photograph C-1. Trail left by the star Betelgeuse taken with a 90mm refractor, F/D = 80 on Kodak 2415 developed in D19. The telescope was fixed. The irregularities are due to atmospheric agitation. This star was chosen for its great brightness for reasons of illustration in this book; a precise measurement of agitation would require a thinner trail, thus a star one or two magnitudes dimmer.

INDEX